Flight
Instructor
Manual

55 Inverness Drive East, Englewood, CO 80112-5498
International Standard Book Number 0-88487-183-5

JS314711D

INTRODUCTION

The *Flight Instructor Manual* is a comprehensive textbook reference designed for commercial pilots who desire a flight instructor certificate. Academic material for the instrument and multi-engine instructor ratings also is provided. An applicant may enroll in the three instructor courses consecutively and become a fully qualified CFI with instrument and multi-engine airplane instructor privileges at the conclusion of the training programs. An applicant who has already acquired a basic flight instructor certificate may acquire either or both of the additional instructor ratings. The manual and other materials are designed for applicants training under FAR Part 141 in approved schools. They also may be used for those applicants training under FAR Part 61.

The manual is divided into three parts: textual information, exercises, and pilot briefings. The textbook section is divided into five chapters, which represent logical divisions of flight instructor subject areas. Corresponding exercises allow applicants to check their progress in the academic phase of the course. At strategic points in the course, pilot briefings provide the basis for in-depth discussions with the instructor covering the academic knowledge and practical skills associated with the flight instructor certificate and ratings.

The manual is supplemented with several other comprehensive study materials.

1. Flight Instructor Stage and Final Examinations are designed to evaluate the applicant's progress throughout the course. The examinations are required items for a flight instructor course conducted under FAR Part 141 in an FAA-approved school. Individual questions are taken from the actual FAA Question Books.

2. The *Flight Instructor Training Syllabus* provides a lesson-by-lesson guide for completion of the ground and flight training required for all applicants who are enrolled in approved schools. It also is beneficial for applicants seeking certification outside of approved schools. The syllabus reflects the content of the FAA Flight Instructor Practical Test Standards.

3. The Flight Instructor File Folder is designed to provide a record of ground and flight training accomplished in the course. This file folder is an important requirement for graduation from approved schools, and it is beneficial for training conducted under FAR Part 61.

4. Flight Instructor Video Presentations introduce and reinforce the concepts presented in the *Flight Instructor Manual*. These motivating presentations are available at participating schools and are designed to enhance and complement other student study materials.

5. The Flight Instructor Courses Insert to the *Courses Instructor's Guide* contains information on the three separate programs — basic instructor, instrument instructor, and multi-engine instructor. Included are answers, where appropriate, for the Pilot Briefings and answer keys for the Flight Instructor Stage and Final Examinations.

TABLE OF CONTENTS

chapter 1

THE BASIC FLIGHT INSTRUCTOR

INTRODUCTION

The material covered in this chapter is appropriate to the applicant for a flight instructor certificate with an airplane category rating and a single-engine land class rating. The first section covers the broad range of flight training activities associated with private and commercial pilot applicants. It also describes procedures for instructing in complex airplanes and conducting biennial flight reviews. The second section reviews and highlights the aerodynamic principles and concepts which are useful to the flight instructor. The entire chapter is operationally oriented toward the practical aspects of flight training. Theory of instruction and fundamentals of learning are presented in Chapter 2.

SECTION A—SINGLE-ENGINE AIRPLANE INSTRUCTION

The material in this section emphasizes those areas which are relevant to the beginning instructor. It is oriented toward the student as well as the instructor and, therefore, sets the tone of the student/instructor interaction which is the basis for effective training. This section begins with the transition to the right seat, continues with a description of the integrated method of flight instruction, and concludes with the private and commercial training areas.

TRANSITION TO THE RIGHT SEAT

The first task during flight training for the instructor applicant is the transition to the right seat. This will be accomplished under the guidance and supervision of a qualified instructor who will occupy the left seat. Checking out in the right seat normally involves some surprise and mild frustration, punctuated with a few humorous comments from the instructor. Caution is advised since there also is a certain amount of danger until adjustments are made to compensate for the changed perspective of the new seating arrangement.

ORIENTATION

During engine starting and runup it will become apparent that the airplane was not really designed for easy operation from the right seat. All of the flight instruments, most of the operating switches, and, in some cases, the engine instruments are located on the left side of the airplane. In addition, the power controls require the use of the left hand. These seemingly minute differences will require the gradual formation of new habit patterns. Typically, there is a tendency to take off on the left side of the runway since previous experience has reinforced the idea that the white line belongs slightly to the right of the pilot's normal position. On initial attempts, it is not unusual to have the left wheel on the edge of the runway before liftoff is achieved. The instructor will provide timely guidance as the transition progresses, as shown in figure 1-1.

PROFICIENCY

After initial orientation, the task of attaining private and commercial pilot proficiency from the right seat will be assigned. This means the entire range of maneuvers and procedures, including basic instrument flight, must be performed to private or commercial practical test standards (PTS). Even for a proficient flight instructor applicant, this normally requires three to five hours of instruction and additional solo practice.

Crosswind and specialty takeoffs and landings require the greatest effort and concentration. These procedures tend to accentuate the change in visual perspective between the left and right seats. It is best to proceed cautiously, and gradually work up to the performance requirements specified in the appropriate PTS.

Fig. 1-1. Transitioning to the Right Seat

PRACTICE INSTRUCTION

Following the transition period, practice instruction periods will be assigned and the character of the CFI course will become evident. The instructor will require the CFI applicant to formulate a lesson plan on a given maneuver or series of maneuvers. During the lesson, the instructor will imitate a typical student pilot whose proficiency level is appropriate to the stage of training assigned. The CFI applicant will be expected to explain the maneuver as the demonstration is being performed with the "student" following through on the controls. The "student" will then perform the maneuver while displaying various faults. The CFI applicant will correct and critique the "student's" performance as the lesson progresses.

Lesson assignments for the instructor course will become progressively more involved, but will follow this general format. The instructor will shift roles throughout the practice lessons, first acting as the student pilot then reverting to the instructor role for a critique of the CFI applicant's performance and progress. The CFI applicant will learn to present a specific lesson through this procedure, complete with preflight and postflight briefings. The discussion which follows lays other important ground work for the beginning instructor and analyzes maneuvers from the viewpoint of typical student difficulties.

INTEGRATED METHOD OF INSTRUCTION

Since its introduction, the integrated method of flight instruction has been accepted as a valuable training technique. Essentially, it means that instruction in the control of the airplane by outside visual references is "integrated" with instruction in the use of the flight instruments to perform the same operations. From the first time a maneuver is introduced, the student is required to perform it by instrument as well as visual references.

The objective of this training is to establish good habit patterns for the use of the flight instruments from the student's first piloting experience. Development of the ability to fly in instrument weather is not the purpose, but it does provide positive transfer to later instrument rating instruction. Benefits of this method of training include building confidence in the flight instruments, precise airplane control, increased operating efficiency, and emergency capability.

CONFIDENCE

Confidence in the use and accuracy of the flight instruments must be developed before anyone can control an airplane by instrument reference. This confidence usually is developed during *basic* attitude instrument instruction such as straight-and-level flight, turns, climbs, and descents. Beginning students will see nothing unusual about the procedure, since they have not yet developed habit patterns for flying.

PRECISE CONTROL

During early experiments with the integrated technique of flight instruction, it was recognized that students trained in this manner were much more precise in their flight maneuvers and operations. This finding applies to all flight operations, not just when flight by reference to instruments is required. Students trained under the integrated concept tend to monitor power settings better and more accurately maintain desired headings, altitudes, and airspeeds. As the habit of monitoring performance by reference to instruments is developed, the student soon begins to make corrections without prompting from the instructor.

OPERATING EFFICIENCY

The performance obtained from an airplane increases noticeably as a pilot becomes more proficient in monitoring and correcting flight attitudes by reference to the instruments. The degree of efficiency attained from an airplane depends directly on how closely the flight and engine instruments are monitored, and how precisely the airplane is controlled with regard to heading, airspeed, and altitude.

EMERGENCY CAPABILITY

The most obvious benefit from integrated flight instruction is the ability to maneuver the airplane for limited periods under favorable circumstances if outside visual reference is lost. This emergency capability should allow noninstrument-rated pilots to extricate themselves from unexpected bad weather conditions and to utilize assistance from ATC in the event they become lost and/or trapped above an overcast.

The success of the integrated method of instruction partially depends on how comfortable the student is when performing maneuvers solely by reference to the flight instruments. It is important that the student realize the purpose of the view-limiting device, or "hood," before its actual use is required in flight. Prior to simulated instrument flight, the instructor should take care to ensure the hood is fitted properly and the student is comfortable.

Another important element involves the instructor's ability to help the student develop judgment and good decision-making skills. The student *cannot* attempt flight into actual instrument conditions without an instrument rating but, at the same time, proficiency is required to adequately maneuver the airplane to get out of IFR weather conditions, if necessary.

SAFETY CONSIDERATIONS

Student safety, both during training and after certification, is dependent upon habit patterns. An effective instructor helps a student develop safety-oriented habit patterns in two ways: first, by teaching safe flight procedures; and second, by setting an example of safety consciousness. This second step is as important as the first since it is common knowledge that students develop the same habit patterns that their instructors have. An example of this is teaching students to always take off on the runway centerline using all of the available runway rather than initiating takeoff at an intersection. Landings should consistently be made on the centerline within the first one-third of the runway. Throughout the training program, the instructor should stress safe operating practices.

COLLISION AVOIDANCE

During flight training, the student is busy learning and the instructor is busy teaching and evaluating. These circumstances may cause both to neglect their collision avoidance responsibilities. To avoid this, instructors must recognize they are ultimately responsible for the safety of flight and, during early training fights, they must bear one hundred percent of the collision avoidance responsibility. Through a constant emphasis on safety, students will learn the proper techniques and gradually begin to exercise their share of this vital responsibility. Then, during solo flights, the students will be prepared to exercise the proper collision avoidance procedures.

Constant effort must be exercised to see and avoid other aircraft, particularly while operating in the vicinity of an airport. Before takeoff, the approach and departure paths should be checked closely since it is the pilot's responsibility to stay well clear of other aircraft. This procedure should be followed at controlled, as well as uncontrolled, airports. The use of anticollision and landing lights during flight in the vicinity of an airport will make an aircraft more clearly visible to others. During climbout, the airplane may be accelerated to cruise climb speed as soon as a safe altitude is reached. The higher speed provides a lower pitch attitude and increased forward visibility. Continuous turns of more than 90° should be avoided while in the traffic pattern.

The instructor and student must continually be aware of blind spots caused by the design of their aircraft. As shown in figure 1-2, the wings and fuselage of an aircraft restrict outside visibility. For this reason, clearing turns are necessary to provide adequate visibility of affected airspace. Prior to beginning flight maneuvers, two 90° clearing turns in opposite directions should be executed. The wing in the direction of intended turn should be

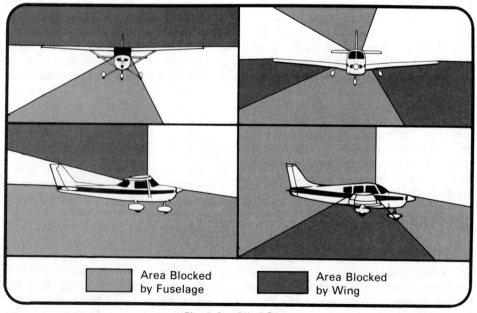

| Area Blocked by Fuselage | Area Blocked by Wing |

Fig. 1-2. Blind Spots

raised or lowered, as appropriate, to provide clear vision of the area before a turn is initiated.

To see something clearly under normal illumination, pilots must look directly at the object. When scanning for other aircraft, the scan should be accomplished by looking at individual sectors, rather than permitting the eyes to sweep across the sky. Each sector of the sky should be brought into focus separately. Students should be taught to spend at least 70% of their time scanning for other aircraft. Early in the training program is the time to help the student develop good scanning habits and overcome the problems associated with empty field myopia, blind spots, and haphazard scan patterns.

Finally, students should be taught to properly use the CTAF as well as all available services, such as the Local Airport Advisory Service (LAA) provided by an FSS, and radar traffic advisories. While students should be encouraged to participate in radar programs for VFR aircraft, they should be cautioned to avoid the false sense of security that often accompanies radar traffic advisories. This is extremely important in Class B and C airspace which receive the heaviest concentrations of air traffic.

WAKE TURBULENCE

Broadly defined, wake turbulence refers to the phenomena that result from the passage of an aircraft through the atmosphere. The term actually includes vortices, thrust stream turbulence, jet engine blast, jet wash, propeller wash, and rotor wash, both on the ground and in the air. Students must become acutely aware of associated hazards. Wingtip vortices are produced, to some degree, by all airplanes developing lift as air flows from the bottom of the wing over the wingtip into the area of low pressure air at the top of the wing. This circulation causes a whirlpool or vortex of air to form behind the tip of each wing. The intensity of the vortices depends on the aircraft weight, speed, and configuration. An airplane generates the greatest wake turbulence when it is heavy,

clean, and slow. Since these conditions primarily exist during takeoff, climbout, and landing, it is within the airport environment that the greatest danger exists. In fact, wingtip vortices from commercial jets can induce uncontrollable roll rates in smaller aircraft. The probability of induced roll is greatest when the small aircraft's heading is aligned with the generating aircraft's flight path.

At the present time, the only reliable way to avoid wake turbulence is to recognize and avoid areas where it is likely to be encountered. Experience has shown that vortex cores tend to descend at 400 to 500 feet per minute behind a landing aircraft. The vortices continue to sink at this rate until they reach approximately 900 feet below the aircraft, where they begin to gradually break up and dissipate. When the vortices are generated in calm wind conditions near the ground, they will settle until within about 200 feet of the ground and then move laterally outward. A crosswind will decrease the lateral movement of the upwind vortex and increase the movement of the downwind vortex. Thus, a light wind of three to seven knots could result in the upwind vortex remaining in the touchdown zone for a period of time and hasten the drift of the downwind vortex toward another runway. Similarly, a tailwind condition can move the vortices of the preceding aircraft forward into the touchdown zone. The student should be warned that the light quartering tailwind requires maximum caution.

The principal rule to remember is that hazardous wake turbulence is generated whenever a large aircraft is airborne. The best avoidance procedure in a small airplane is to remain *above* the wake turbulence. The pilot of a light aircraft should land beyond the touchdown point of a large airplane and take off well before the liftoff point of a large airplane. When landing after a departing large aircraft, the small aircraft should touch down on the approach end of the runway as far as practical from the rotation point of the departing airplane. When landing abeam a large aircraft on parallel runways, the small airplane should remain above the flight path of the large one and land beyond its touchdown point.

JET BLAST

Every year, the number of ground accidents involving serious injury and aircraft damage due to jet blast indicate that some pilots are not aware of this serious hazard. No pilot would deliberately taxi an aircraft in a hurricane, yet some have taxied behind a jet generating winds of hurricane force. As shown in figure 1-3, the winds generated by jet engines at takeoff thrust can reach velocities of 300 m.p.h. The table in the lower part of the illustration shows the distances required to avoid jet blast velocities greater than 25 m.p.h. Even when the generating aircraft is at idle thrust, as much as 620 feet may be needed to avoid dangerous jet blast velocities. Jet blast can be avoided by holding well back of the runway edge when waiting for takeoff and by aligning the aircraft to face the blast, if possible. If a light aircraft must be taxied behind a large jet, ground control should be requested to inform the jet pilot not to increase thrust above idle until the light aircraft is well clear.

MICROBURSTS AND WIND SHEAR

Microburst and low-level wind shear are additional hazards that students need to be made aware of during training. A microburst is an intense, localized downdraft which spreads out in all directions when it reaches the surface. This creates severe horizontal and vertical wind shears which pose serious hazards to aircraft, particularly those near the surface. Any convective cloud can produce this phenomenon. Although microbursts are commonly associated with heavy precipitation in thunderstorms, they often are associated with virga. Virga is the term used for streamers of precipitation that trail beneath a cloud but evaporate before they reach the ground. When there is little or no precipitation at the surface, a ring of blowing dust may be the only evidence of a microburst.

The microburst downdraft is typically less than one mile in diameter as it descends from the cloud base to about 1,000 to 3,000 feet AGL. In the transition zone near the ground, the downdraft changes to a horizontal outflow that can extend to approximately two and one-half miles in diameter, with peak winds lasting only two to five minutes. The downdrafts can be as

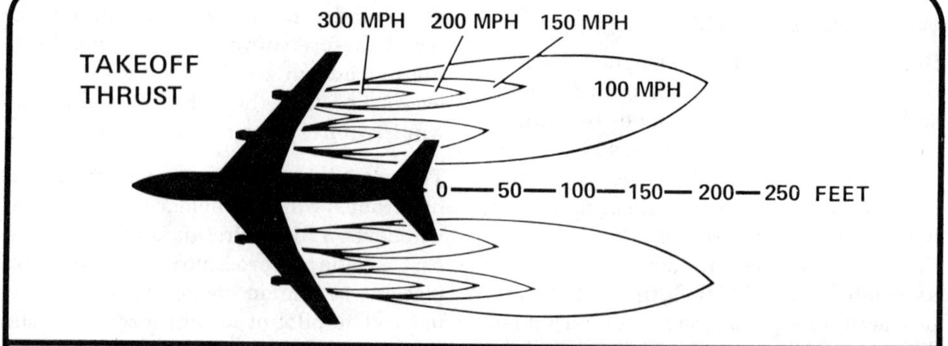

25 MPH VELOCITY	B-727	DC-8	DC-10/L-1011	B-747, B-767, B-757, A-300, and all re-engined B-707/DC-8 aircraft
Takeoff Thrust	550 ft.	700 ft.	2,100 ft.	1,752 ft.
Breakaway Thrust	200 ft.	400 ft.	850 ft.	1,250 ft.
Idle Thrust	150 ft.	100 ft.	350 ft.	620 ft.

Fig. 1-3. Avoiding Jet Blast

strong as 6,000 f.p.m., and the horizontal winds near the surface can be as strong as 45 knots. The horizontal wind shear can result in a sudden change in indicated airspeed. The amount of change is directly related to how fast the wind speed or direction changes. As shown in figure 1-4, if an aircraft inadvertently takes off into a microburst, it first experiences a headwind which increases performance, item 1. This is followed rapidly by decreased performance as it encounters the downdraft, item 2, and the wind shears to a tailwind, item 3. This may result in terrain impact or operation dangerously close to the ground, item 4. An individual microburst will seldom last longer than 15 minutes from the time it strikes the ground until it dissipates. The horizontal winds continue to increase during the first five minutes, with the maximum intensity winds lasting approximately two to four minutes. Sometimes microbursts are concentrated into a line structure. Under these conditions, activity may continue for as long as an hour. Once microburst activity starts, multiple microbursts in the same general area are not uncommon and should be expected.

INSTRUCTING THE PRIMARY STUDENT

Most private pilot courses are designed to accommodate four general areas of instruction in addition to final practical test preparation. The general categories are shown in the following list.

1. Basic airwork
2. Low altitude airwork
3. Traffic pattern
4. Cross-country and night flights

The obvious advantage of this sequence is that it represents a logical, step-by-step, building block progression of learning which makes the job easier and more effective for both student and instructor. The following paragraphs discuss this general sequence of training.

BASIC AIRWORK

Basic airwork includes maneuvers such as straight and level, turns, climbs, descents, maneuvering at critically slow airspeed, and stalls. Normally, they must be completed no lower than 1,500 feet AGL, which is in accordance with the practical test standards (PTS). These maneuvers may well be the most important phase of training because they form the basis for the development of all piloting skills. All flight maneuvers, regardless of how complex, are only variations or combinations of straight and level, turns, climbs, and descents. Every student *must* have a thorough understanding of these four basic maneuvers and be able to

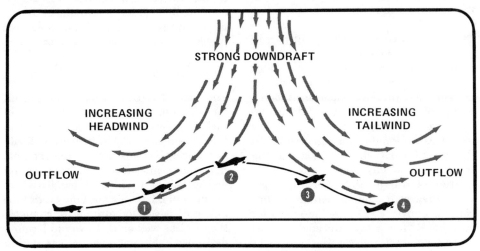

Fig. 1-4. Microburst on Departure

perform them accurately before proceeding to other maneuvers and procedures.

ORIENTATION

The orientation flight requires good organization and planning on the part of the instructor. The purpose of the flight is to give the student an introduction to flying, and a broad overview of the training program. It is not designed to cram the student full of the specific details of flying. Instructors should remember the potential student may not be completely convinced that learning to fly is the best course of action, and a confusing orientation flight can dampen motivation or even cause a student to reconsider the whole idea.

The best course of action during the orientation flight is for the instructor to encourage the student to relax and enjoy the experience. The student should follow through on the controls and experiment with them at altitude while the instructor points out distant landmarks. Many instructors include a short cross-country during the orientation flight. The total effect on the student should be relief and enthusiasm. After the flight, students normally are quite vocal with questions if they have decided to go through with flight training. The instructor should respond candidly to the student's inquiries, then advise them what to expect during the next few lessons.

The second lesson, which is still part of the orientation process, gives the instructor an opportunity to introduce a detailed preflight inspection, emphasize the use of the checklist, and lay important groundwork for future training. One of the important things to accomplish in the first few lessons is to formulate an agreement with the student about who will be controlling the airplane at any given time. Usually instructors will assert, "You've got it," when they want the student to assume control. Conversely, when the instructor commands "I've got it," the student must understand that the controls are to be relinquished immediately.

STRAIGHT AND LEVEL

Frequently, a student's difficulty with complex maneuvers can be traced to a lack of understanding of the proper techniques for straight-and-level flight. The lesson usually begins with a demonstration of the visual and instrument references associated with straight-and-level flight. The proper attitude for straight-and-level flight is established and the relative position of the nose and wing tips to the horizon is described. The demonstration continues with the instructor making deviations from straight and level and pointing out the changes in both visual and instrument references.

As the pitch attitude is changed, the airspeed variation may be emphasized as a secondary indication. This is especially important when the student is using the attitude indicator as the principle reference for straight-and-level flight. The correct indications for each flight instrument should be explained by the instructor, first while the student is using outside references and, again, when only instrument references are utilized.

In the initial stage of training, one problem the integrated method of instruction may create is the tendency for the student to depend on the instruments alone for attitude information. The instructor should continually look for indications of overreliance on the instruments and correct the problem before it becomes a habit.

TRIM CONTROL

The concept of trim normally is difficult for beginning students since nothing in their previous experience relates well to this type of control. The instructor should devote considerable time to the explanation and demonstration of trim control during initial training. If students are required to retrim the airplane every time airspeed, power, or attitude is changed, they will be well on their way to forming desirable habit patterns for use during their entire flying careers.

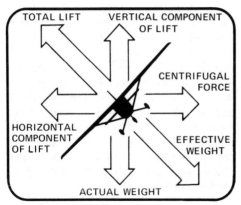

Fig. 1-5. Forces in a Turn

STEEP TURNS

A thorough explanation is beneficial before the introduction of steep turns. The student should understand the forces which cause an airplane to turn and the proper function of the rudder during the turn. Figure 1-5 illustrates the distribution of lift and weight components in a turn. The instructor should emphasize that only the direction of the total lift vector changes during the turn. The ailerons are used to change the direction of the lift vector, the rudder is used to keep the turn coordinated, and the elevator or stabilator is used to maintain altitude.

The instructor should include an explanation of the aileron pressure required during the various turns. For training purposes, turns are considered either shallow, medium, or steep. The shallow bank is one in which the stability of the airplane must be overcome by maintaining slight control pressure in the direction of turn. In a medium bank turn, the control pressures are balanced, which requires neutralization of the ailerons after the bank is established. Steep turns require slight aileron pressure opposite the direction of turn to oppose the overbanking tendency.

Initially, lack of coordination is the major problem during turn entries and recoveries. Students often apply rudder pressure either too soon or too late. Coordination problems also result when a student tries to sit up straight during a turn, rather than riding with the airplane.

Consistent loss of altitude during turns is another common student problem that results from the failure to compensate for the loss of vertical lift with additional elevator back pressure. If a student loses altitude during a left turn but gains altitude during a right turn, the cause usually is a change of perspective. A dot properly placed on the windshield in front of the student may alleviate this problem, since it provides the same reference for both left and right turns.

For the private pilot practical test, steep turns of 360° must be demonstrated using a bank angle of 45° (±5°) while maintaining coordinated flight. A constant altitude (±100 feet) and airspeed (±10 knots) must be maintained and the student must roll out of the turn within 10° of the desired heading. The steep turn must be performed in either direction as specified by the examiner. In addition, the student must exhibit knowledge of the performance factors associated with steep turns, including increased load factors, power required, and the overbanking tendency.

CLIMBS AND DESCENTS

The effect of power changes on pitch and yaw usually is demonstrated before the introduction of climbs and descents. An effective method of demonstration is to have the student maintain straight-and-level flight while the instructor first decreases power, then increases power to maximum. The effects of torque and P-factor can be shown in the same manner. The instructor can have the student raise the nose of the airplane to a climb attitude while at cruise power with the feet off of the rudders. The instructor should define the climb performance desired in terms of attitude and power before practice begins. During initial practice, a specific level-off altitude is not important until the student is comfortable and has the feel of a stabilized climb.

Descents should be practiced initially at the power and airspeed recommended by the airplane manufacturer. At first, the descents can be prolonged to allow the student to become accustomed to the required pitch attitude and control pressures. Gliding descents with the power off at the best glide speed should be introduced *after* partial power descents have been demonstrated with an acceptable level of performance. If prolonged gliding descents are practiced, it may be necessary to clear the engine by periodically advancing the throttle to ensure proper engine operation.

As soon as basic proficiency is gained in straight climbs and descents, climbing and descending turns should be introduced. They may be entered from stabilized straight climbs or descents, or from straight-and-level flight. In the latter case, the pitch and bank attitudes must be established simultaneously, and it is easier for the student to check for other traffic while the climb or descent is being established.

Common errors during climb and descent maneuvers include failure to cross-check and correctly interpret outside and instrument references. Other problems may be the application of control movements rather than pressures, poor correction for torque effect, and improper use of trim. Precise airspeed and pitch control during climbs and descents are prerequisites for takeoffs and climbs as well as approaches and landings.

MANEUVERING DURING SLOW FLIGHT

Other terms such as "flight at critically slow airspeed" or "flight at minimum controllable airspeed" have been used when referring to "slow flight." The latter is the term used in the Private and Commercial Pilot PTS. Recognition of the flight characteristics at speeds just above a stall and precise airspeed control are the objectives of maneuvering during slow flight. If a pilot is able to recognize the sounds and visual cues associated with an airplane at slow airspeeds, the chance of entering a stall inadvertently is decreased. Practice of slow flight will help the student operate

safely at the slower airspeeds characteristic of takeoffs, departures, and landing approaches.

During the demonstration of the maneuver at altitude, the instructor should point out the decreased sound level and the loss of control effectiveness which results from reduced velocity of the relative airflow. Particular attention should be given to coordination because of the large rudder displacement needed to overcome left-turning tendencies.

When stabilized in straight-and-level flight at the specified slow flight airspeed, gentle turns of approximately 10° of bank usually are practiced to demonstrate the additional power required to prevent altitude loss. Steeper turns should be practiced after the applicant is reasonably proficient in making shallow turns in both climbs and descents. In addition, the student should practice the maneuver in various configurations while the instructor emphasizes the changes in flight characteristics that occur.

Some of the common student errors include improper entry technique and failure to establish and maintain slow flight airspeed. Other problems include the inability to establish the specified configuration and excessive variations in altitude and heading from that specified. See the appropriate PTS for performance criteria.

STALLS

Preparation for the introduction of stalls should begin with a ground discussion and review of what a stall is and why it occurs since stalls can be a source of apprehension to student pilots. The discussion should include a description of a stall and its recovery so the applicant will know what to expect. The actual instruction in stalls should begin with an explanation of the cues which indicate an impending stall. The practice of slow flight should have acquainted the student with these cues, but they should be reviewed and reinforced with the introduction of stalls.

Even though an airplane can be stalled in *any* attitude, at *any* airspeed, and with *any* power setting, inadvertent stalls are most likely to occur with a nose-high attitude. The unusually high pitch attitude is the most obvious visual cue. Other indications of an impending stall are the decrease in tone and volume of the sound of the air and the rapid decay of control effectiveness.

The first few stalls should be entered from a normal glide with the power off. Recovery should be initiated at the first physical indication of a stall, which normally is the onset of aerodynamic buffeting or decay of control effectiveness. Recoveries should always be made with power to establish a habit pattern of positive response to a potential stall situation. Full stalls are practiced with back elevator or stabilator pressure held until the nose "breaks," immediately followed with recovery procedures. Proper stall recovery normally involves three items performed as a simultaneous response.

1. Reducing the angle of attack by relaxing back pressure
2. Adding power to conserve altitude
3. Leveling the wings to regain a normal flight attitude

If the stall is being practiced in the approach or landing configuration, flaps should be retracted to the takeoff setting or position recommended by the manufacturer.

Full stalls should be practiced with the power on and off, in straight and turning flight, and in configurations appropriate for both takeoff and departure and approach and landing. Entry altitude must allow the recoveries to be completed no lower than 1,500 feet AGL. Students can expect to demonstrate these stalls for the practical test.

There are a number of common student errors associated with stalls including failure to establish the proper configuration prior to entry. Other problems include improper pitch, heading, or bank control during straight-ahead stalls, improper pitch or bank control during turning stalls, and rough or uncoordinated control techniques. Additional items include improper compensation for torque during power-on stalls, poor stall recognition and delayed recovery, excessive altitude loss or excessive speed during recovery, and development of a secondary stall.

STALL/SPIN AWARENESS

While stall/spin awareness has always been an important part of pilot training, regulatory changes to FAR Part 61 and Part 141 now require additional ground and flight training in this area. These regulations affect both aeronautical knowledge and flight proficiency requirements for recreational, private, and commercial pilot applicants who are seeking airplane or glider category ratings. The additional training required should include approximately one hour of ground instruction and one hour of flight instruction. Of course, this time will vary with the individual student and other circumstances in the training environment.

GROUND TRAINING

A good reference for this subject is AC 61-67B, *Stall and Spin Awareness Training*. It explains the stall and spin awareness training required under FAR Part 61 and offers guidance to flight instructors who provide that training.

RECOMMENDED SUBJECTS

The ground training subjects recommended by the FAA are as follows:

1. Definitions
   a. Angle of attack
   b. Airspeed
   c. Configuration
   d. V-speeds

e. Load factor
f. Center of gravity (CG)
g. Weight
h. Altitude and temperature
i. Snow, ice or frost on the wings
j. Turbulence
2. Distractions
3. Stall Recognition
4. Types of Stalls
   a. Power-off (Approach to Landing)
   b. Power-on (Departure Stalls)
   c. Accelerated
5. Stall Recovery
6. Secondary Stalls
7. Spins
   a. Weight and balance
   b. Primary cause
8. Types of spins
   a. Incipient spin
   b. Fully developed spin
   c. Flat spin
9. Spin Recovery

INADVERTENT STALLS

The primary cause of an inadvertent spin is exceeding the critical angle of attack for a given stall speed while executing a turn with excessive or insufficient rudder and, to a lesser extent, aileron. In an uncoordinated maneuver, the pitot/static instruments, especially the altimeter and airspeed indicator, are unreliable due to the uneven distribution of air pressure over the fuselage. The pilot may not be aware that a critical angle of attack has been exceeded until the stall warning device activates. If a stall recovery is not promptly initiated, the airplane is more likely to enter an inadvertent spin. The spin that occurs from cross controlling an aircraft usually results in rotation in the direction of the rudder being applied, regardless of which wing tip is raised. In a skidding turn, where both aileron and rudder are applied in the same direction, rotation will be in the direction the controls are applied. However, in a slipping turn, where opposite aileron is held against the rudder, the resultant spin will usually occur in the direction opposite the aileron that is being applied.

FLIGHT TRAINING

The flight training requirements reflect the findings of the FAA's *General Aviation Pilot Stall Awareness Training Study*. This study showed that instructors can educate students about the traps that lead to an unintentional stall and provide them with practice in avoiding unintentional stalls when challenged by the distractions that often cause stall accidents. The study also concluded that this combination of ground and flight training was very effective in reducing the occurrence of unintentional stalls and spins. The intention of the regulations is for flight instructors to emphasize recognition of situations that could lead to an inadvertent stall/spin.

REALISTIC DISTRACTIONS

The FAA recommends that flight instructors simulate scenarios that can lead to inadvertent stalls by creating distractions while the student is practicing certain maneuvers. The idea is to distract the student from his or her primary objective — maintaining aircraft control. This can be done by giving the student a task to perform while flying at slow airspeeds. Actually, there are any number of distracting tasks that can be assigned. They include the following:

1. Retrieving a dropped pen or pencil
2. Determining a heading to an airport using a chart
3. Resetting the clock to Coordinated Universal Time (UTC)
4. Retrieving something from the back seat
5. Interpreting the outside air temperature gauge
6. Calling FSS for weather information
7. Computing TAS with a flight computer
8. Identifying terrain or objects on the ground
9. Identifying a suitable emergency landing site
10. Alternately climbing/descending 200 feet and maintaining altitude
11. Reversing course after a series of S-turns
12. Flying at slow airspeeds with the airspeed indicator covered

STALL AVOIDANCE AT SLOW AIRSPEEDS

The practice of slow flight is a very effective stall avoidance maneuver. A heading and altitude should be assigned and the student should fly slow enough to keep the stall warning device activated. The left-turning tendencies caused by torque and P-factor should be demonstrated to show why heavy right rudder is required. Experiment with elevator and rudder trim to show their effects at slow speeds and also demonstrate adverse yaw. In addition, students can make turns, climbs, and descents to improve their control at slow speeds. Flap extension and retraction at slow speeds is another beneficial item for student practice. This exercise can be combined with another difficult task — flying with the airspeed indicator covered. According to AC 61-67B, stall demonstrations and practice, including maneuvering during slow flight and other maneuvers with distractions that can lead to inadvertent stalls, should be conducted at a sufficient altitude to enable recovery above 1,500 feet AGL in single-engine airplanes and 3,000 feet AGL in multi-engine airplanes.

STALLS AFTER TAKEOFF

Another exercise for practice includes stalls after takeoff which can be simulated at a safe altitude. Students should be instructed to perform these straight ahead and in turns while in coordinated flight. Then have the students repeat the stalls with feet off the rudder pedals to demonstrate the effects of poor coordination. Emphasize the tendency to spin when the stall occurs in uncoordinated flight. Finally, a distraction can be provided just prior to the stall, and any effects it may have had on the recovery can be explained. For the student's benefit, power-on (departure) stall practice should be related to the takeoff phase of flight.

ENGINE FAILURE/180° TURN

While emphasizing how an inadvertent stall might actually occur on takeoff, a simulated engine failure at altitude in a climb can be demonstrated, followed by a 180° turn. Students will quickly see just how must altitude is lost when an attempt is made to return to the airport following an engine failure in the takeoff phase. They should also see how rapidly airspeed decays when an engine fails on takeoff and why an immediate pitch down adjustment is needed to maintain best glide speed.

CROSS-CONTROLLED STALLS

Another good demonstration exercise is performance of cross-controlled stalls in gliding turns. This situation simulates turns from base to final and permits the student to observe the effects of coordinated flight, as well as slips and skids, on stall and stall recovery. The ball position in the inclinometer should be pointed out so the student understands what is happening during slips and skids when the airplane stalls.

POWER-OFF STALLS

Power-off (approach-to-landing) stalls and go-arounds offer additional avenues for developing stall and spin awareness in students. After they have performed a recovery from a full-flap, power-off stall, the total loss of altitude should be pointed out and related to the traffic pattern altitude. The student should be instructed to recover without retracting flaps and then attempt to climb. It is very likely that a secondary stall will occur if a climb attitude is maintained. Also, the student can be shown what can happen if the flaps are retracted too quickly. Emphasize how critical a configuration error can be during stall recovery at low altitudes during an approach.

STALLS DURING GO-AROUNDS

Stalls during go-arounds continue to be cited in accident reports. To avoid this situation, have students perform a full-flap, gear-extended, power-off stall at altitude. Then have them recover and attempt to climb with flaps extended. If a higher than normal climb pitch attitude is held, a secondary stall will result. In fact, a stall will occur in many airplanes in a normal climb pitch attitude. Another variation following recovery is to have students retract the flaps rapidly while holding a higher than normal climb pitch attitude. A secondary stall or setting with a loss of altitude usually will result. These procedures should heighten the student's awareness of the stall potential during go-arounds.

ELEVATOR TRIM STALLS

Another useful demonstration stall involving simulated go-arounds is an elevator trim stall. First, the aircraft is placed in an approach configuration at altitude and trimmed for a descent. After the descent is stabilized, a go-around should be initiated while allowing the nose to pitch up and torque to swerve the airplane to the left. At the first indication of a stall, a recovery to a normal climbing pitch attitude should be initiated. Students must understand why correct attitude control, application of control pressures, and proper trim are so important during go-arounds and why incremental retraction of flaps is a must. Students will benefit from these demonstrations when they begin concentrated takeoff and landing practice in the traffic pattern.

LOW ALTITUDE AIRWORK

Ground reference maneuvers prepare students for flight operations involving the observation or use of ground objects such as those necessary for approaches and landings and traffic pattern practice. Each of the ground reference maneuvers require students to divide their attention between flying the airplane and following prescribed paths over the ground. The altitude at which ground reference maneuvers are performed depends on the airspeed of the training airplane. An altitude between 600 and 1,000 feet above all obstructions usually is adequate, but the following factors should be considered when choosing the altitude.

1. The groundspeed should not be so fast that events happen too rapidly.
2. The path of the airplane over the ground should be obvious.

3. Drift should be easily discernible, but it should not tax the student's ability to make corrections.
4. The altitude should be low enough so any appreciable change in altitude is obvious.

DETERMINING WIND DIRECTION

Prior to actually performing any ground reference maneuver, wind direction and velocity should be ascertained. There are several easy methods that can be demonstrated to the applicant. The most common method of determining wind direction is to observe a column of smoke or dust, and note the direction it is blowing. The velocity of the wind can be estimated by the angle the dust or smoke column makes with the ground. If smoke or dust are not visible, the surface of a lake or river can be used to estimate wind conditions.

A wind drift circle may be used to determine the wind when there are no obvious indications outside the airplane. To fly a wind drift circle, the airplane is flown over a prominent ground reference point. When directly over the point, a 360° turn is entered and the bank angle is maintained throughout the turn. At the completion of the turn, the direction and distance from the point where the bank was entered indicate the effect of wind during the turn. Figure 1-6 illustrates the effect of various wind conditions on a wind drift circle.

Another method of determining wind direction and velocity is to fly a straight line between two points and observe the drift correction required. This method tells a pilot only what approximate wind correction angle is required for the particular course being flown; not the actual wind direction or velocity. However, the maneuver is a very effective introduction to rectangular courses, and it is effective for showing the student how to compensate for wind.

RECTANGULAR COURSES

Rectangular courses introduce the techniques used in traffic patterns. The objec-

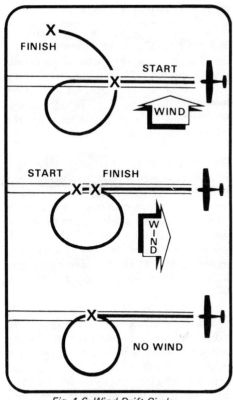

Fig. 1-6. Wind Drift Circles

tive of the rectangular course is to maintain a specified distance from a rectangular pattern on the ground while maintain-

ing a specific altitude. To accomplish this, the student must understand the effect of wind and how to correct for it in both straight and turning flight. Practice of rectangular courses develops skill in maneuvering the airplane with a minimum amount of attention to attitude or instrument indications. Figure 1-7 illustrates the correct performance of rectangular courses.

Common errors in the performance of rectangular courses include poor planning, orientation, or division of attention, uncoordinated flight control application, and improper correction for wind drift. Additional problems include failure to maintain selected altitude or airspeed and selection of a ground reference where there is no suitable emergency landing area within gliding distance.

S-TURNS ACROSS A ROAD

While demonstrating S-turns, the instructor must be certain the student understands all factors involved — the changing bank angles required with changing groundspeed, the variation of back pressure necessary due to varying bank angles, and the requirement for accurate coordination. Incorrect adjustment of bank angle to compensate for changing groundspeed

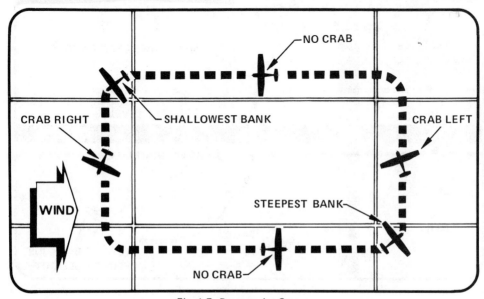

Fig. 1-7. Rectangular Course

causes unequal patterns on opposite sides of the road. This problem occurs when a student flies the pattern mechanically without correcting for wind. Therefore, it is necessary for the maneuver to be flown in changing wind conditions to provide practice in wind drift correction. Figure 1-8 illustrates the varying bank angles during the execution of an S-turn.

Inaccurate scanning and incorrect use of back pressure may result in poor altitude control. In addition, when operating close to the ground, there is often a tendency to hold elevator or stabilator back pressure, causing a gain in altitude and a loss of airspeed. This problem normally stems from the student's unconscious effort to maintain maximum ground clearance.

The instructor also should be alert to the student's coordination during the turns and while rolling from one bank to the other. If problems are encountered, returning to altitude to practice coordination might be valuable. It is very important that basic flying techniques and coordination not be ignored or allowed to deteriorate during the practice of ground reference maneuvers.

TURNS AROUND A POINT

Turns around a point usually are entered on a downwind heading. This entry procedure simplifies memorization of the details of each ground reference maneuver, since most are entered in this fashion. After proficiency is attained, however, the student should be encouraged to enter the maneuver at any point.

The most common problems associated with turns around a point involve incorrect drift correction resulting in oval shaped patterns, or circles with constantly varying radii. Since many students are unable to visualize a correct track over the ground, it is important that the instructor identify problems early in the training so the maneuver is practiced correctly by the student.

A conscious effort may be required to return the student's attention to the instruments when it has been focused on a ground reference point; therefore, instrument scanning techniques are developed by the practice of this maneuver. Coordination, planning, and timing also are improved as skill is acquired. The increased attention required outside the aircraft

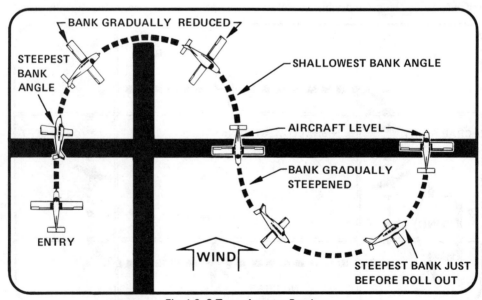

Fig. 1-8. S-Turns Across a Road

during turns around a point requires the student to fly the airplane by feel rather than mechanical reactions. Common errors in the performance of turns around a point and S-turns include faulty entry techniques; poor planning, orientation, or lack of division of attention; and improper correction for wind drift.

All of the maneuvers discussed under low altitude airwork develop the student's ability to plan and execute a maneuver by reference to an object on the ground while compensating for wind. Always caution the student to maintain at least minimum obstruction clearance during the performance of ground reference maneuvers and to be sure a suitable landing area is within gliding distance. Performance tolerances for these maneuvers include maintaining altitude within 100 feet and airspeed within 10 knots.

TRAFFIC PATTERN

Traffic pattern procedures require the student to apply all the maneuvers and techniques learned in earlier training. The student should be familiar with the legs of the traffic pattern by this stage of the training program. In addition, the student should have made at least a few unassisted takeoffs by the time concentrated traffic pattern practice is begun.

NORMAL TAKEOFFS AND CLIMBS

The takeoff is one of the easiest maneuvers to teach if the proper techniques have been introduced. Taxiing experience has a direct bearing on the student's initial takeoff performance since steering, power changes, and proper control positioning for wind also are involved during the takeoff. The instructor should emphasize crosswind takeoff procedures at every opportunity during formal traffic pattern practice. This will provide a positive carryover to the more difficult landing phase of training.

Directional control problems on takeoff usually can be traced to inadequate taxi techniques or the use of improper visual cues during the takeoff run. Normally, the student's vision should be concentrated to the left side and midway down the runway when the takeoff is initiated. As the takeoff continues, the student's visual perspective should be shifted progressively farther down the runway.

The student should be required to begin rotation gradually and lift the nosewheel off the runway slowly, rather than accelerating in a three-point stance followed by abrupt rotation. After rotation, the student should be required to set up the proper crab angle for climbout on the extended runway centerline. Turns after takeoff should be shallow, and they should be initiated only when within 300 feet of traffic pattern altitude.

Although takeoff practice in a tricycle-type airplane is relatively easy, the instructor must remain alert for potential student errors or sudden faltering of the engine. This is also an opportunity to establish a well-ingrained habit on the part of the student for making a check of engine power instruments after full throttle is applied on takeoff. Many instructors prefer their students to reply verbally that the power check has been completed and that the airspeed indicator is functioning.

After reasonable proficiency in takeoffs is attained, it is beneficial for the instructor to simulate engine failure on takeoff by suddenly retarding the throttle. This lays the groundwork for conditioning the student to the possibility of an actual engine failure and provides exposure to the sudden transition from acceleration to deceleration. If runway length permits, the power can be reduced at progressively later stages in the takeoff run and even after liftoff, provided the student has gained reasonable proficiency in landings. This will provide practice in adjusting rapidly from a climb to a descent.

NORMAL APPROACHES AND LANDINGS

Although beginning instructors seldom find anything "normal" about the student pilot's early landing attempts, they will find this phase of training to be one of the most interesting. It often requires great patience and understanding on the part of the instructor but, most importantly, it requires keen observation and analysis. Instructors should begin approaches and landings with the full realization that the student knows only what has been taught in previous dual sessions. If the instructor has laid a solid groundwork in basic maneuvers, the student should achieve steady progress in the landing phase.

On the other hand, the early solo flight still receives too great an emphasis among student pilots and many instructors. A premature solo is very dangerous; therefore, from the beginning of training, the student should be aware that solo will occur only when the instructor feels that it can be performed *safely*. If possible, initial instruction in landings should be conducted at an airport with a small volume of air traffic to allow consistent approach patterns and a minimum of distractions. With fewer variables in the initial approaches, the student can more easily visualize and establish the final approach path.

STABILIZED APPROACH

An approach that utilizes a specific airspeed and power setting and, therefore, a constant rate of descent is a "stabilized approach." Each type of landing may require a different airspeed or approach angle for a given configuration; however, the parameters of airspeed and power setting can be defined for any approach such as a soft-field or short-field approach.

Whatever the method used, the value of a stabilized approach cannot be over emphasized. It is very difficult to execute a good landing while constantly attempting to adjust the glide path; therefore, all maneuvering should be accomplished as early as possible on the final approach. The student's goal should be to establish the correct airspeed and glide path during the first portion of the final approach, using only minor corrections during the last segment.

The instructor should have introduced landings by having the student follow through on the controls during the first few flights. It will be easier for the student if each variable during the approach and landing can be minimized when formal instruction in approaches and landings actually begins. A constant airspeed and power setting for each leg of the approach should be chosen. The initial power reduction on downwind should occur when abeam the intended touchdown point followed by a smooth transition to final approach airspeed, with a turn to base leg at the 45° point. The student should soon learn to lead the turn to final so runway alignment is be achieved at roll-out. On final, it is a matter of maintaining approach angle, runway alignment, and adjusting power as necessary to maintain the desired approach speed.

In the past, many instructors preferred to delay the use of flaps until after solo. This eliminated one more variable for the student and also had advantages in the case of a go-around. In addition, since flaps tend to complicate crosswind landings, instructors often felt their use made the landing phase more difficult for the beginning student pilot. Now, it is a generally accepted practice to include the use of flaps during presolo training. In addition, the use of flaps may be required on the practical test so the practice of flap landings is appropriate from the beginning of traffic pattern training.

The instructor should direct the student's vision to the appropriate points as the approach progresses once concentrated landing practice begins. Vision is obviously the most important sense used in landing practice, and considerable evidence shows that visual concentration on the wrong areas during an approach causes bad landings and a great deal of student/instructor frustration. Instructors can observe their students' visual patterns or simply ask them where they are looking. Beginning students have difficulty judging their position in relation to the desired approach path during final approach. Since the student has no background experience in this area, the instructor must carefully demonstrate the stabilized approach, emphasizing a constant approach angle and pointing out appropriate visual cues.

Instructors have developed a variety of theories concerning the correct visual cues and general runway perspective that a student should observe during a stabilized approach. Techniques which are effective for one instructor are seldom in agreement with those of another. The value of any particular technique is dependent upon the student's progress in achieving consistent approaches and landings.

As the approach progresses, the student's vision should shift to the intended touchdown point. Figure 1-9 illustrates that, if a constant approach angle is maintained and the airplane is aimed at the desired touchdown point, that point will appear to remain stationary.

During the flare, students should be prompted to shift their vision to the left and further down the runway edge since the nose will tend to block out the runway as

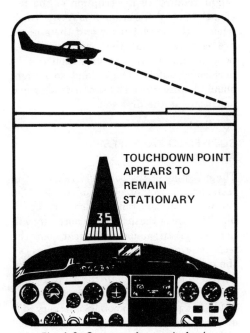

Fig. 1-9. Constant Approach Angle

TOUCHDOWN POINT APPEARS TO REMAIN STATIONARY

35

the flare progresses. Depth perception problems may occur if the student's vision is not focused the correct distance ahead of the airplane. Inconsistency in flaring altitude can be the result of changing the focal point during the approach. To alleviate this problem, the student should be encouraged to focus the same distance ahead of the airplane as in a car traveling at the same speed.

CROSSWIND TAKEOFFS AND LANDINGS

Ideally, instructors prefer student pilots to have their initial lessons in the traffic pattern without wind, followed by several sessions with good crosswind conditions. Usually, the opposite happens, and the student's first attempts in formal traffic pattern training are aggravated by variable crosswinds.

The crosswind takeoff should be demonstrated by the instructor with the student following through on the controls. The student should be taught to begin the takeoff with full upwind aileron and maintain runway alignment with rudder. As the airplane accelerates, control pressures can gradually be reduced.

In strong crosswinds, the use of the aileron control may cause the downwind wheel to lift off the runway first and may result in the completion of the takeoff roll on one wheel for some distance. If the proper amount of aileron is used to correct for the existing crosswind, and the airplane is held on the ground until a positive liftoff can be made, there should be no unusual side load on the landing gear. The student should be required to transition smoothly to the required crab angle after liftoff and climb straight out on the extended runway centerline.

A large part of a student's problems with crosswind takeoffs and landings stems from the fact that they are really cross-control maneuvers. All previous training has emphasized coordinated control usage as a primary objective; therefore, it will take some time to develop the concept of crosswind control technique. Extensive preflight and postflight briefings will be beneficial to the student.

The crosswind landing requires the student to set up the proper crab angle on final and transition to a side slip during or just prior to the flare. This is most easily accomplished by simultaneously applying rudder to align the airplane with the runway while applying aileron pressure to lower the upwind wing. Touchdown should occur on the upwind main gear first, followed by the downwind main gear and nose gear, respectively. The student must maintain the proper amount of upwind aileron during the roll-out and exert enough steady forward pressure on the control wheel to hold the airplane firmly on the runway. These procedures will require a great deal of concentration and practice for an individual with only a few hours of flight experience.

During initial crosswind landing practice, many instructors have their students lower the upwind wing and apply opposite rudder on long final. This stabilizes the approach and simplifies the flare and touchdown for the student. The technique allows the student to determine the amount of aileron and rudder pressure required. The instructor should periodically demonstrate the correct procedure during the lesson to keep the student oriented and to help break the tension the student may experience.

The student must realize that a good crosswind landing requires alignment of the airplane's longitudinal axis, flight path, and the runway at touchdown. In some cases, it will be necessary to return to the practice area for more fundamentals or concentrated practice transitioning from a crab to a slip.

The instructor should not forget to solicit the student's analysis of a particular problem. This procedure often reveals a detail that has been overlooked or an erroneous concept the student has adopted during training. Crosswind takeoffs and landings present peculiar problems to both student and instructor, but they also provide great satisfaction and confidence once they are mastered.

FIRST SOLO

When ready to solo, the student will have consistently demonstrated the ability to proficiently perform all of the fundamental maneuvers. In addition, the student should be capable of handling ordinary problems that might occur in the traffic pattern, such as conflicting traffic or a change in the active runway caused by variable winds. For this reason, the student should have been exposed to crosswinds during the practice of takeoffs and landings and should have demonstrated adequate proficiency in moderate crosswinds.

Ordinarily, the student should be required to make three landings to a full stop during the first solo flight. The instructor then has the opportunity to stop the flight if circumstances warrant. Unusual traffic volume, gusty winds, or poor performance might require an interruption of the lesson. Generally, the student is ready to solo when both the instructor *and* the student are confident that the preparation has been adequate and the operation can be performed safely. The second solo flight should be supervised by the instructor just as closely as the first solo.

SOFT-FIELD TAKEOFFS AND LANDINGS

The objective of the soft-field takeoff is to transfer the weight of the airplane from the wheels to the wings as soon as possible. Practice of these takeoffs normally will not be on an actual soft field, but the *technique* can be simulated regardless of the type of runway surface. A nonstop taxi with a rolling turn onto the runway will simulate the technique used for the softest field that permits takeoff. Some very soft surfaces require the runup to be completed either prior to or during taxi.

As power is applied and the takeoff roll is initiated, full elevator or stabilator back pressure is applied to raise the nosewheel from the soft runway surface. As the speed increases and the elevator or stabilator becomes more effective, a slight reduction in back pressure is required to maintain the necessary constant-pitch attitude. If the back pressure is not reduced, the airplane may assume an extremely nose-high attitude which, in some airplanes, can cause the tail skid to come in contact with the runway or cause premature liftoff.

If the proper pitch attitude is maintained, liftoff will occur at a lower-than-normal airspeed. As the airplane lifts from the runway surface, a further reduction in back pressure must be initiated to achieve a level flight attitude. The airplane then must be accelerated in level flight, within ground effect, to the best angle-of-climb airspeed before the flaps are retracted. If the flaps are retracted prematurely, the resultant loss in lift may cause the airplane to settle back onto the runway.

Soft-field approaches are essentially the same as normal approaches. A small amount of power is carried during the flare and touchdown to keep the airplane's weight off of the wheels as long as possible. If the landing is executed properly, the total weight of the airplane is transferred slowly from the wings to the wheels. The nosewheel should remain clear of the surface during most of the landing roll. Directional control is very important during the after landing roll. Figure 1-10 illustrates the proper technique for soft-field takeoffs and landings.

SHORT-FIELD TAKEOFFS AND LANDINGS

Short-field takeoff procedures must utilize the best combination of takeoff and climb performance. Since runway length and obstructions are of primary importance, it is obvious that each takeoff should take advantage of the entire length of the runway. The manufacturer's recommendations for airplane configuration, rotation, and climb speed must be observed.

General technique should be taught so that later transition to other airplanes is easier. For example, every short-field takeoff, regardless of airplane type, should start from the very end of the runway with smooth application of full power. The airplane should be rotated for takeoff at the recommended speed. Then, the best angle-of-climb speed (or obstacle clearance speed) should be maintained until all obstacles are cleared.

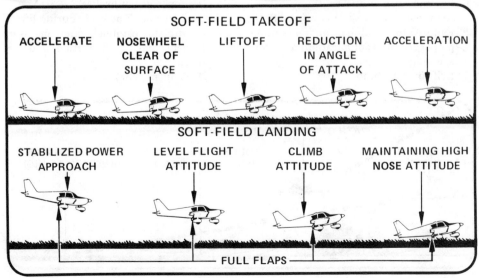

Fig. 1-10. Soft-Field Takeoff and Landing

During a short-field landing, it is assumed the approach and landing are made over a 50-foot obstacle. The landing is performed from a full-flap, stabilized-power approach with the touchdown executed in a power-off, full-stall attitude. By incorporating this technique, the airplane can be maneuvered safely and accurately over the obstacle to a landing touchdown at the slowest possible groundspeed, producing the shortest possible ground roll. Figure 1-11 illustrates how an approach that is too shallow or too fast will cause an increase in the landing distance.

COMMON ERRORS

There are a number of common student errors in the performance of takeoffs and landings. The following listings are grouped under the type of operation.

Normal and Crosswind Takeoffs and Climbs

1. Improper use of takeoff and climb performance data
2. Improper initial positioning of flight controls or wing flaps
3. Improper power application
4. Inappropriate removal of the hand from the throttle
5. Poor directional control
6. Improper use of aileron
7. Improper pitch attitude during liftoff
8. Failure to establish and maintain proper climb configuration and airspeed
9. Drift during climb

Normal and Crosswind Approaches and Landings

1. Improper use of landing performance data and limitations
2. Failure to establish approach and landing configuration at the proper time or in proper sequence
3. Failure to establish and maintain a stabilized approach
4. Failure to use the proper technique for wind shear or turbulence
5. Inappropriate removal of the hand from the throttle
6. Faulty technique during roundout and touchdown
7. Poor directional control after touchdown
8. Improper use of brakes (landplane)

Short-Field or Maximum Performance Takeoff and Climb

1. Improper use of takeoff and climb performance data
2. Failure to position the airplane for maximum utilization of the available takeoff area
3. Improper initial positioning of flight controls or wing flaps
4. Improper power application
5. Inappropriate removal of the hand from the throttle
6. Poor directional control
7. Improper use of brakes
8. Improper pitch attitude during liftoff
9. Failure to establish and maintain proper climb configuration and airspeed
10. Drift during climb

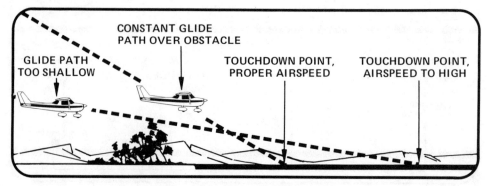

CONSTANT GLIDE PATH OVER OBSTACLE

GLIDE PATH TOO SHALLOW

TOUCHDOWN POINT, PROPER AIRSPEED

TOUCHDOWN POINT, AIRSPEED TO HIGH

Fig. 1-11.   Short-Field Approach

Short-Field or Maximum Performance Approaches and Landings

1. Improper use of landing performance data and limitations
2. Failure to establish approach and landing configuration at the proper time or in the proper sequence
3. Failure to establish and maintain a stabilized approach
4. Faulty technique in the use of power, wing flaps, and trim
5. Failure to use the proper technique for wind shear or turbulence
6. Inappropriate removal of the hand from the throttle
7. Faulty technique during roundout and touchdown
8. Poor directional control after touchdown
9. Improper use of brakes

Soft-Field Takeoff and Climb

1. Improper initial positioning of the flight controls or wing flaps
2. The hazards of allowing the airplane to stop on the takeoff surface prior to initiating takeoff
3. Improper power application
4. Inappropriate removal of the hand from the throttle
5. Poor directional control
6. Improper use of brakes
7. Improper pitch attitude during liftoff
8. The hazards of settling back to takeoff surface after becoming airborne
9. Failure to establish and maintain proper climb configuration and airspeed
10. Drift during climb

Soft-Field Approach and Landing

1. Improper use of landing performance data and limitations
2. Failure to establish approach and landing configuration at the proper time or in the proper sequence
3. Failure to establish and maintain a stabilized approach
4. Failure to consider the effect of wind and landing surface
5. Faulty technique in the use of power, wing flaps, or trim

6. Failure to use the proper technique for wind shear or turbulence
7. Inappropriate removal of the hand from the throttle
8. Faulty technique during roundout and touchdown
9. Failure to hold back elevator pressure after touchdown
10. Closing the throttle too soon after touchdown
11. Poor directional control after touchdown
12. Improper use of brakes

GO-AROUNDS

Students should be instructed in go-arounds from rejected or balked landings from the beginning of takeoff and landing practice. Two important elements are the student's ability to recognize the need for a go-around and the necessity of making a timely decision to do so. Other performance items include the use of recommended airspeeds, being aware of the drag effect of wing flaps and landing gear, and being able to properly cope with undesirable pitch and yaw tendencies. There are a number of reasons for go-arounds, such as a bounced landing, sudden gusty winds, unexpected traffic on the runway, or the inability to achieve touchdown on the first third of the runway.

Common student errors related to go-arounds include failure to recognize a situation where a go-around is necessary or delaying a decision to make a go-around. Other items include improper power application and failure to control pitch attitude or compensate for torque effect. Some students may improperly retract the wing flaps or landing gear. Others may fail to maintain the proper track on climbout and remain clear of obstructions and other traffic.

EMERGENCY LANDINGS

Practice emergency landings are designed to simulate an engine failure during some segment of a flight. The simulation should be introduced with sufficient altitude to allow the student to formulate and execute a plan of action. As the training progresses, the simulations should contain an element of surprise to add realism to the practice. A simulated engine failure or partial power loss should occur during high and low altitude airwork and just after takeoff. In addition to preparing the student for forced landings, the procedure enables the instructor to evaluate the student's judgment and ability to think and react under stress. Performance is evaluated on the basis of promptly establishing the best glide speed and the recommended configuration. Other factors include selection of a suitable emergency landing area, use of the emergency checklist, and the importance of attempting to determine the cause of the malfunction.

The most common student errors during simulated emergency approaches and landings include improper airspeed control, poor judgment in the selection of an emergency landing area, and failure to estimate the approximate wind speed and direction. Additional items include failure to accomplish the emergency checklist or to fly the most suitable pattern for the existing situation as well as under-shooting or over-shooting the selected emergency landing area. When errors are detected during practice that might affect the outcome of the approach, the instructor should prompt the student to make immediate corrections.

CROSS COUNTRY

The cross-country phase of a training program builds confidence and experience. This will be the first time students can use their new skills for a practical way. Dual cross-country flights also provide a break in the maneuvers routine for the instructor and the student.

Before the first dual cross country, a briefing session with the sectional chart is an absolute necessity. Accurate navigation cannot be expected if the student is not familiar with how to use the chart. There cannot be enough emphasis placed on the importance of knowing the chart thoroughly before takeoff.

CHOOSING THE ROUTE

Route selection is one of the most difficult tasks the instructor faces in the cross-country phase of training. Each cross-country route should be chosen to provide the student with the largest variety of experience possible. The cross-country route is normally triangular-shaped with each of the three legs more than 50 nautical miles in length. The dual cross-country should have one leg that is common to a leg of the first solo cross-country route. This will reduce student apprehension about the first solo cross-country flight. The points of landing should be chosen to challenge the student and to provide experience in landing at different types of airports.

PREFLIGHT PLANNING

Preflight planning is a very important element of any flight and, especially, cross-country flights. From the standpoint of safety, this is a good place to help the student establish good habit patterns for preflight preparation. The important elements of preflight planning involve checking the flight information publications, aviation weather reports and forecasts, and determining airplane performance, including computation of weight and balance and fuel requirements.

Before flights into unfamiliar areas, most experienced pilots determine boundaries and course limits. It may be a good idea for the instructor to point this out to the student in the cross-country stage of training. Figure 1-12 illustrates how this technique can be used. For example, during the leg from Platte to Chamberlain, a powerline can be used as the *right course boundary* and the river for the *left course boundary*. The *course limit*, in this case, is the bend in the river and the road north of the city of Chamberlain. Limits and boundaries make it easier for the student to recognize that a course correction is necessary.

The completion of a navigation log normally is the last item of preflight planning.

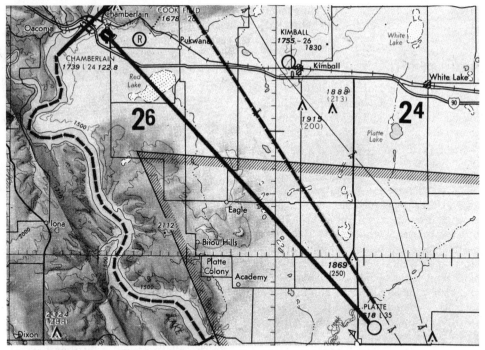

Fig. 1-12. Course Boundary and Limit

Students typically have difficulty choosing good checkpoints which are appropriate to a particular route so the instructor should provide guidance in this area. Requiring a completed navigation log for each cross-country reinforces good habit patterns. The information in the navigation log is crucial to the student during flights over unfamiliar terrain. Instructors should carefully inspect the completed navigation log following solo cross-country flights to determine the accuracy of the student's estimates.

THE FLIGHT

The first dual cross-country provides the instructor with an opportunity to introduce pilotage, dead reckoning, and radio navigation in one instruction period. If the flight has been planned with three separate legs, each leg can be flown using a different type of navigation. The distinction between pilotage and dead reckoning should be made on this flight, regardless of the predominant method of navigation.

Techniques for unfamiliar airport operation also should be covered thoroughly during the dual cross-country phase. The first step is preflight planning. Most of what a pilot needs to know about an unfamiliar airport can be found in the *Airport/Facility Directory* or determined from the sectional chart. Overflying the airport at a safe altitude above the traffic pattern to observe traffic, obstructions, and wind direction also is a recommended procedure.

The instructor should point out to the student that some uncontrolled and many controlled airports have nonstandard traffic patterns. At controlled fields, the solution is simply to follow instructions from the controller. Nonstandard traffic patterns at uncontrolled fields usually are noted in the *Airport/Facility Directory* or described by the airport advisory service or UNICOM operator if one is available. In any case, it is a recommended practice to monitor the CTAF from at least 10 miles out during letdown to any uncontrolled airport.

Each dual cross-country should include a hypothetical lost situation. The instructor should quiz each applicant to determine what action would be taken in the event of disorientation. If a direction finding facility is available, a practice DF steer also should be obtained.

NIGHT FLIGHT

A ground session dealing with the distinctive aspects of night operations, including airport markings and lighting, should be completed prior to the first night flight. The instructor will need to describe the process of dark adaptation and the limitations of vision at night including the possibility of disorientation and optical illusions. Other items include how to conduct a visual inspection at night and the importance of a clean windshield. The use of position and anticollision lights prior to starting as well as the proper adjustment of cockpit lighting also are appropriate areas for emphasis. During the flight, the instructor should include night emergencies and approaches and landings with and without landing lights.

The flight should include a thorough familiarization with the local area. Distances at night are deceptive because of an inability to compare the sizes of different objects. More dependence must be placed on the flight instruments, especially the altimeter and airspeed indicator. The importance of carrying an operable flashlight (preferably with a red lens) should be stressed. Simulation of instrument and cabin light failures on the first night flight will convince the applicant.

Outside light is probably the most important factor for a night takeoff. If there is no moon or stars and the takeoff is made in a direction away from a city or toward an area where there are no lights, the principal features of an instrument takeoff are duplicated. In these situations, a complete transition to instrument reference is required.

The most common errors associated with night approaches and landings probably will be the student's tendency to flare too late and fly the airplane into the runway. A power approach is sometimes the best solution for this problem. Students should

be trained to use power during the final segment of the approach and "feel" for the runway. Regardless of the technique used, the student should realize the importance of maintaining currency if regular night operations are planned. The instructor should also discuss the various landing illusions that can be experienced in the night environment.

The instructor should emphasize to the student that the following factors must always be considered before making a night flight.

1. Visibility
2. Amount of outside light available
3. Surface winds
4. General weather situation
5. Availability of lighted airports enroute (cross-country flight)
6. Proper functioning of the airplane and its systems
7. Night flying equipment in the airplane
8. The pilot's recent night flying experience

SYSTEMS AND EQUIPMENT MALFUNCTIONS

Each system or equipment malfunction creates its own distinct set of problems. Although, an instructor cannot simulate every possible malfunction, the student can be conditioned to react calmly and rationally in adverse conditions and maintain positive control of the airplane. In addition, the student should become familiar with the recommended action specified in the pilot's operating handbook and follow the appropriate emergency procedure checklist. The following list contains typical systems and equipment malfunctions. Depending on the aircraft, other malfunctions are possible.

1. Smoke or fire, or both, during ground or flight operations
2. Rough running engine or partial power loss
3. Loss of engine oil pressure
4. Fuel starvation
5. Engine overheat
6. Hydraulic system malfunction
7. Electrical system malfunction

8. Carburetor or induction icing
9. Door or window opening in flight
10. Inoperative or "runaway" trim
11. Landing gear or flap malfunction
12. Brake failure
13. Any other system or equipment malfunction

Recommended action for most of these items varies with the manufacturer although some general rules apply. For example, any engine fire in flight requires the mixture to be set at idle cutoff and the fuel selector turned off followed by an emergency landing. Fuel starvation is another item that happens to many pilots each year. Students should be made aware that a precautionary landing with low fuel is preferred to a forced landing due to fuel exhaustion.

Procedures for a partial or complete power loss should follow the airplane manufacturer's recommendations. The instructor should require the applicant to use the emergency checklist as a guide. A recommended method is to start with the fuel quantity gauges, select the fullest fuel tank, and turn on the electric fuel pump. Carburetor heat then can be applied and the mixture enriched. The final items that should be checked are the magnetos. Occasionally, a mis-timed magneto will cause a partial power loss. When the defective magneto is isolated from the ignition system, some power may be restored. Total engine failure is rare, but each student should be prepared to deal with the problem if it does occur.

EMERGENCY DESCENT

Most POHs include emergency descent procedures for incidents such as an in-flight fire, cloud penetration, or sudden loss of pressurization. The instructor should emphasize the importance of following the emergency descent procedure listed in the POH for the specific emergency. Emergency descent considerations may include high drag configurations (gear and flaps down), the use of carburetor heat, and observing specific V-speed and aircraft structural limitations. An emergency de-

scent may involve fairly high airspeeds and positive G loading.

EMERGENCY EQUIPMENT AND SURVIVAL GEAR

For cross-country flights it is wise to carry emergency equipment and survival gear. This equipment should be stowed in a location that can be quickly accessed after an emergency landing. Usually emergency equipment and survival gear is stowed in the aft portion of the aircraft where it has the greatest chance of remaining intact after a forward crash. When stowing the material, emphasize to your students that all emergency equipment and survival gear should be secured so it is not tossed about the cabin in turbulence or during an off-airport landing.

INSTRUCTING ADVANCED MANEUVERS

The advanced maneuvers listed in the Commercial Pilot PTS require a high degree of pilot proficiency. Many of these maneuvers are performed near the airplane's performance limits.

STEEP TURNS

Steep turns aid in the development of coordination, planning, and control use while operating near the performance limits of the airplane. The major emphasis in the performance of the maneuver should be placed on proper pitch and bank control. Control pressures continually change as bank angle is increased during roll-in. When the applicant is established in the turn, the control pressures remain fairly constant.

Steep turns should be accomplished with the primary emphasis on visual references. There may be a tendency, however, for the applicant to concentrate on the flight instruments, since they provide more specific information concerning pitch and bank attitude.

Steep turns of 360° are performed at bank angles of 50° with a 5° tolerance. Roll-out should be within 5° of the specified heading immediately followed by at least a 360° turn in the opposite direction. Altitude should not vary more than 100 feet and a constant airspeed must be maintained within 5 knots. Common student errors include improper pitch, bank, and power coordination during roll-in and roll-out as well as uncoordinated or inappropriate use of flight controls. Other items include improper technique in correcting altitude deviations, loss of orientation, and exceeding tolerances on the roll-out heading. Because of the high load factors imposed, the student should be cautioned to remain below maneuvering speed. In addition, any indication of a stall must be avoided. Finally, instructors should emphasize the difference in visual perspective between left and right turns to help the student maintain altitude as shown in figure 1-13.

CHANDELLES

A chandelle is a maximum performance 180° climbing turn incorporating a smooth transition from the entry speed to within a few knots above the stalling speed. The maneuver requires a high degree of advanced planning, accuracy, coordination, smoothness, and awareness of control responses. A chandelle, when properly practiced, develops good coordination habits and control sensitivity awareness due to the continuously changing pitch attitudes, airspeeds, and control pressures.

Like the steep turn, the chandelle is conducted mainly by use of outside visual references, so the instructor must be sure the applicant is using these references. The chandelle is illustrated in figure 1-14. Common student errors include improper pitch, bank, and power coordination during entry or completion, and uncoordinated use of flight controls. Additional items include improper planning and timing of pitch and bank attitude changes or other factors related to failure in achieving maximum performance, stalling during the maneuver, and excessive deviation from the desired heading during completion.

LAZY EIGHTS

Lazy eights require top performance in planning, timing, and coordination because pitch and roll attitude, heading, and airspeed change constantly. The lazy eight

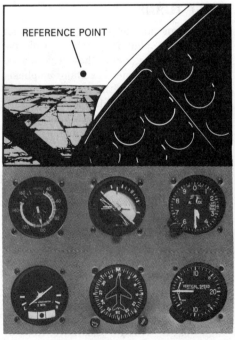

Fig. 1-13. Steep Turn Reference Points

consists of two 180° turns in opposite directions, each containing a gentle climb and a gentle dive. The maneuver is called a lazy eight because the longitudinal axis of the aircraft appears to scribe a flight pattern about the horizon that resembles a figure eight lying on its side.

Prominent reference points should be selected on the horizon as an aid to making symmetrical loops during each turn. The reference points should be 45°, 90°, and 135° from the direction in which the maneuver is begun. Figure 1-15 illustrates the flight path and a plan view of a lazy eight.

The maneuver is entered from straight-and-level flight at cruise or maneuvering speed. A very shallow turn is established as the nose is raised gently to a climb attitude. At the 45° point in the turn, pitch attitude should be at its highest, while bank angle is increasing to one-half of its maximum of 30°. From the 45° point to the 90° point, pitch attitude slowly returns to level while bank angle continues to increase. At the 90° point, the bank angle is at its maximum and the nose passes through level flight toward the descent attitude. At the 135° point, the nose is at

its lowest, and the bank angle decreases about 15°. At the 180° point, the airplane passes through straight-and-level flight, progressing toward an identical turn in the opposite direction. Altitude and airspeed should be the same as they were at entry.

Some common errors during performance of lazy eights are included in the following list.

1. Poor selection of reference points
2. Uncoordinated use of flight controls
3. Unsymmetrical loops resulting from poorly planned pitch-and-bank attitude changes
4. Inconsistent airspeed and altitude at key points
5. Loss of orientation
6. Excessive deviation from the reference points

The lazy eight is almost as difficult to evaluate as it is to perform, since it is a gently flowing maneuver with no factor remaining constant. However, progress can be checked every 45° of turn. Evaluation must be made on the basis of the attitude at each point, as well as coordination, orientation, planning, division of attention, and positive, accurate control.

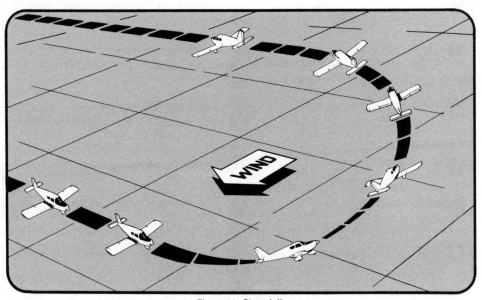

Fig. 1-14. Chandelle

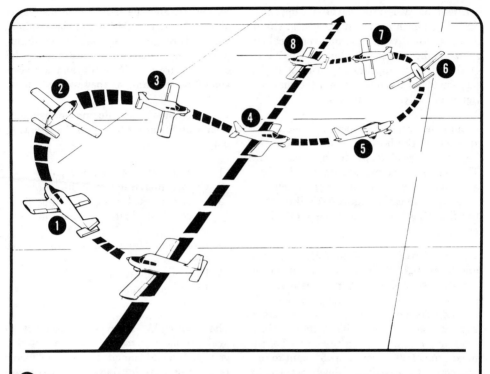

1 0° - 45° Pitch attitude is increased to its highest point while bank is gradually increased to one-half of the maximum.

2 45° - 90° Pitch attitude is gradually decreased to pass through the level flight attitude as the 90° point is reached while the bank is increased to its maximum.

3 90° - 135° Pitch attitude continues to decrease to its lowest point while bank is gradually decreased to one-half of the maximum.

4 135° - 180° Pitch attitude is gradually increased to the level flight attitude as the 180° point is reached while the wings are rolled to level. The airplane is in straight-and-level flight.

5 180° - 225° Pitch attitude is again gradually increased to its highest point at the 225° point while the direction of turn is reversed and the bank is increased to one-half of its maximum.

6 225° - 270° Pitch attitude is decreased to the level flight attitude while the bank is increased to its maximum.

7 270° - 315° Pitch attitude is continually decreased to its lowest point while bank is gradually decreased to one-half of the maximum.

8 315° - 360° Pitch attitude is increased to the level flight attitude while the wings are rolled to level. The airplane is again in straight-and-level flight.

Fig. 1-15. Lazy Eight

ELEMENTARY EIGHTS

Elementary eights involve the same principles of wind drift correction as turns about a point except that a change in the direction of flight is required. Elementary eights can be flown along a road, across a road, or around two pylons. The length of the straight portion of the "legs" can be varied, or the radius of the turn altered for diversity.

Introduction of elementary eights may begin with eights along a road. This maneuver is based on a single ground reference which is usually a straight segment of a road or a section line. The objective is to fly circular-shaped ground tracks on either side of the reference line in a moderate wind condition. Eights across a road is a maneuver that involves extending the principles learned during eights along a road. However, the wind should be perpendicular to the road for this maneuver. Accurate performance of eights-around-pylons requires considerable proficiency in the fundamentals of ground reference maneuvers. The objective is to fly symmetrical patterns about two reference points on the ground.

These maneuvers usually are introduced and practiced by entering on a downwind, or nearly downwind heading for the sake of uniformity. After the maneuvers have been mastered, the student should be able to enter on any heading and perform satisfactorily.

EIGHTS-ON-PYLONS

This required commercial maneuver involves flying circular paths in the form of a figure eight around two points, or pylons, on the ground. With eights-on-pylons, the student does not attempt to maintain a uniform radius around the pylon. Instead, the airplane is flown at an altitude and airspeed combination that makes the wingtip appear to pivot on the point, or pylon, on the ground. This altitude is called the pivotal altitude.

Instruction in pylon eights should begin with the selection of a reference point on the airplane. The reference point should be located near the wingtip on a line extending from the student's eye parallel to the lateral axis of the airplane. The distance of the reference point above or below the wingtip is dependent on the student's height and how near the student's eye level is to the wing while sitting in the aircraft.

Before the maneuver is demonstrated in the airplane, the student must understand pivotal altitude and its relationship to groundspeed so the instructor should plan an extensive preflight briefing. The student must realize that the AGL height of pivotal altitude is governed by the aircraft's groundspeed. As groundspeed increases, the pivotal altitude also increases. An estimate of pivotal altitude AGL may be computed during preflight with the formula:

$$\frac{(TAS \text{ in m.p.h.})^2}{15*} = \text{pivotal altitude}$$

* Use 11.3 for knots

At 100 knots (115 m.p.h.), pivotal altitude will be about 882 feet AGL. Since the exact TAS, groundspeed, and elevation seldom are known, precise pivotal altitude is best determined by experimentation. The best way to do this is to have the student select a point on the ground and begin a circling descent from an altitude which is higher than the estimated pivotal altitude. Descending from above pivotal altitude to a point below pivotal altitude causes the apparent motion of the reference point to slow, stop, and then move forward. The point at which the motion stops is the pivotal altitude. After the student is able to consistently maintain the line-of-sight reference on the selected point, two pylons may be selected to begin practice of the actual maneuver.

Since this maneuver is performed at a relatively low altitude, pylons should be selected in open areas and not on the lee side of hills or near obstructions. Pylons that are at or near the same elevation should be selected to avoid the added burden of adjusting altitude for variations in terrain. The selected pylons should be along a line which is 90° to the direction of the wind, and spaced to provide three to five seconds of straight-and-level flight between the turns, as shown in figure 1-16.

The student must realize that variations in groundspeed will affect pivotal altitude. Throughout each turn, altitude must gradually but constantly, be adjusted to hold the reference point on the pylon. The aircraft at positions 1 and 4 in figure 1-16 are at the roll-in points for the respective turns. At positions 2 and 5, the groundspeed is the slowest; therefore, bank is shallowest, distance is greatest, and the pivotal altitude is the lowest. As the aircraft proceeds to positions 3 and 6, groundspeed increases so bank is steepest, distance is closest, and the pivotal altitude is highest. At positions 3 and 6, the aircraft is rolled into straight-and-level flight for three to five seconds.

Movement of the line-of-sight reference is the student's best indication that pivotal altitude is not being maintained. The left side of figure 1-17 shows that, if the reference point moves ahead of the pylon, the student needs to increase the altitude slightly. The student should soon realize that a climb has a double effect since it also decreases groundspeed which lowers

the pivotal altitude required to maintain the pylon. Climbing above pivotal altitude is shown on the right side of figure 1-17. In this situation, the student needs to descend in order to maintain the pylon. This has a double effect since the descent tends to increase airspeed and raise pivotal altitude.

Common student errors during eights-on-pylons include faulty entry technique, poor planning, orientation, and division of attention. Other items include uncoordinated flight control application, use of an improper line-of-sight reference, application of rudder alone to maintain the pylon, improper timing of turn entries and roll-outs, and improper correction for wind drift.

COMPLEX AIRPLANE

FAR Part 61.129 requires that each applicant for a commercial pilot certificate receive at least 10 hours of instruction and practice in an airplane with retractable landing gear, flaps, and a controllable-pitch propeller. The instructor should remember that this regulation defines an *experience* requirement for the commercial pilot certificate which should not be confused with the limitations placed on both private and commercial pilots by FAR Part 61.31. According to 61.31, a person holding a private or commercial pilot certificate may not act as pilot in command of an

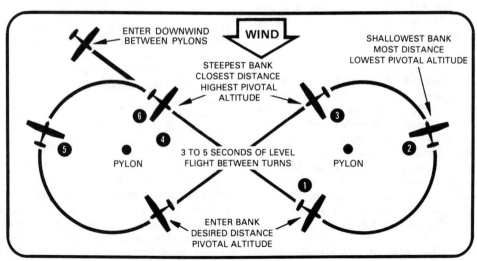

Fig. 1-16. Eights-on-Pylons

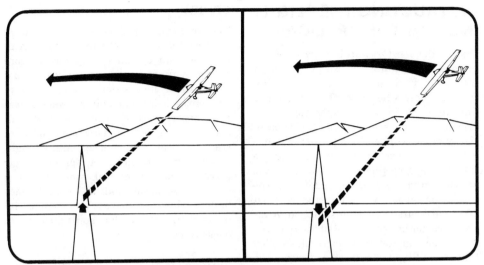

Fig. 1-17. Holding the Pylon

airplane that has more than 200 horsepower, or retractable landing gear, flaps, and a controllable propeller unless that person has received the appropriate instruction from an authorized flight instructor who certifies in the person's logbook that they are competent to pilot such an airplane.

Complex airplane instruction for commercial pilot applicants or high performance airplane checkouts, as described in FAR Part 61.31, should begin with familiarization of the pilot's operating handbook followed by a preflight briefing. Every applicant should have a thorough understanding of each airplane system and its operation before the first flight. In addition, performance should be reviewed with regard to takeoff and landing distances, V-speeds, normal operating altitudes, and loading prior to takeoff.

During the visual inspection, the instructor should identify any unfamiliar items. These items are usually the propeller control, cowl flaps, and landing gear system. In addition, most complex or high-performance airplanes have more advanced fuel systems than light training aircraft, and any unique characteristics should be pointed out.

Most of the problems encountered during the transition to these airplanes are the

result of the increased speed and complexity of the aircraft. At first, the applicant will probably be "behind" the airplane mentally. This usually results in an increase in traffic pattern size with the downwind leg extended and long, low, final approaches. When the applicant gains some familiarity with the airplane, the size of the traffic pattern normally decreases and more normal approaches will be demonstrated. It should be pointed out to the applicant that the proper airspeed control becomes increasingly important as airplane size and speeds increase.

The use of a printed checklist also becomes increasingly important for safe operations. Checklist procedures should be mandatory during any training activity because they increase safety and form valuable habit patterns for use during a pilot's entire flying career. For example, by requiring an applicant to use the printed prelanding checklist, the possibility of a gear-up landing is reduced during initial solo operations. The instructor must impress upon the pilot that use of the printed checklist should become a permanent procedure, not just a requirement for training.

INSTRUCTING THE FLIGHT INSTRUCTOR APPLICANT

One of the most rewarding areas of the profession is training the prospective flight instructor. FARs place experience limitations on those who are authorized to give *flight* instruction to instructor applicants. The instruction required for flight proficiency must be given by a person who has held a flight instructor certificate during the 24 months preceding the date the instruction is given. In addition, the instructor must have conducted at least 200 total hours of flight instruction. After an instructor has accumulated the required instructor experience, it is recommended that the Flight Instructor Course be reviewed in its entirety. This will permit the experienced instructor to become more closely oriented to the needs of the instructor applicant.

DEMONSTRATION STALLS

One of the unique training areas for flight instructor applicants includes demonstration stalls. These are stalls which a flight instructor demonstrates to students to enhance their understanding of stall/spin awareness. They include the following:

1. Cross-control stalls
2. Elevator trim stalls
3. Secondary stalls

On the CFI practical test, applicants must exhibit instructional knowledge of these types of stalls by demonstrating and simultaneously explaining them. An important part of these demonstrations is the selection of the appropriate landing gear and flap configurations. The idea behind these demonstrations is to expose the student to situations where unintentional stalls and spins could develop. The student will be better prepared to recognize potential stall/spin situations and also know how to initiate immediate corrective action. However, students must understand that they should not practice demonstration stalls during solo operations. Refer to the Flight Instructor PTS for specific performance criteria for CFI applicants.

SPINS

The performance of spins is required by FAR 61.183 for a flight instructor certificate with an airplane or glider category rating. A flight instructor applicant must demonstrate instructional competency in spin entry, spins, and spin recovery unless the inspector conducting the test will accept a logbook endorsement of spin training and competence. Many flight instructors also introduce spins to their private and commercial students as a safety precaution and as a confidence builder. However, private and commercial pilots should not conduct spins during solo operations.

According to AC 61-67B, "A spin in a small airplane or glider is a controlled or uncontrolled maneuver in which the glider or airplane descends in a helical path while flying at an angle of attack greater than the angle of maximum lift. Spins result from aggravated stalls in either a slip or a skid. If a stall does not occur, a spin cannot occur. In a stall, one wing will often drop before the other and the nose will yaw in the direction of the low wing."

When introducing spins to flight instructor applicants, it is also helpful to discuss the types of spins. The *incipient spin* is that portion of a spin from the time the airplane stalls and rotation starts, until the spin is fully developed. Incipient spins that are not allowed to develop into a steady state spin are commonly used as an introduction to spin training and recovery techniques. A *fully developed spin* means the angular rotation rates, airspeed, and vertical speed are stabilized from turn to turn and the flight path is close to vertical. A *flat spin* is characterized by a near level pitch and roll attitude with the spin axis near the CG of the airplane. Recovery from a flat spin may be extremely difficult or even impossible.

In a spin, the airplane is pulled downward by gravity, rolling and yawing in a spiral path. It has been estimated that several hundred factors actually contribute to spinning. From this, it is evident that, whether or not spinning is a desirable maneuver or characteristic, it is a feature of airplanes that must be reckoned with in the training of pilots.

Spin demonstrations and practice for flight instructor applicants should include spins in both directions. During spin practice, an entry altitude should be selected which allows recovery well above the minimum 1,500 feet AGL required by FAR Part 91.303. For specific entry recommendations, the pilot's operating handbook should be consulted.

Entry to a spin is accomplished from a full, power-off stall. The elevator or stabilator control should be held as far back as possible for the duration of the maneuver. Slightly before or during the instant of stall, full rudder should be applied and maintained in the direction of the desired spin. The ailerons should not be used, but carburetor heat is recommended because of the rapid engine cooling associated with spins.

In some light airplanes it is advisable to continue a certain amount of power during initial spin entry. This serves two purposes: first, it helps to establish spin rotation upon entry by providing increased propeller slipstream on the vertical stabilator; and second, it helps keep the engine running smoothly during the transition from normal power to idling. In some airplanes that are otherwise difficult to spin, a short burst of power serves the first purpose and prevents an unintentional spiral from developing in place of a spin. The throttle should be closed promptly as the spin develops.

According to AC 61-67B, "Before flying an aircraft in which spins are to be conducted, the pilot should be familiar with the operating characteristics and standard operating procedures, including spin recovery techniques, specified in the approved AFM or POH. The first step in recovering from an upright spin is to close the throttle completely to eliminate power and mini-mize the loss of altitude. If the particular aircraft spin recovery techniques are not known, the next step is to neutralize the ailerons, determine the direction of the turn, and apply full opposite rudder. When the rotation slows, briskly move the elevator control forward to approximately the neutral position. Some aircraft require merely a relaxation of back pressure; others require full forward elevator control pressure. Forward movement of the elevator control will decrease the angle of attack. Once the stall is broken, the spinning will stop. Neutralize the rudder when the spinning stops to avoid entering a spin in the opposite direction. When the rudder is neutralized, gradually apply enough aft elevator pressure to return to level flight. Too much or abrupt aft elevator pressure and/or application of rudder and ailerons during the recovery can result in a secondary stall and possibly another spin. If the spin is being performed in an airplane, the engine will sometimes stop developing power due to centrifugal force acting on the fuel in the airplane's tanks causing fuel interruption. It is, therefore, recommended to assume that power is not available when practicing spin recovery. As a rough estimate, an altitude loss of approximately 500 feet per each 3-second turn can be expected in most small aircraft in which spins are authorized. Greater losses can be expected at higher density altitudes."

LIMITATIONS

Many airplanes are prohibited from spin maneuvers. For example, airplanes certified by the FAA in the normal category are prohibited from spins. This is also true of some airplanes in the utility category. There are a number of training airplanes that are approved for spins. Usually, these aircraft are certificated in both normal and utility category, and they may be spun when they comply with utility category weight and CG limitations. Utility category aircraft typically are certified for limited acrobatic maneuvers. For example, maneuvers which may involve more than 60° of bank, including steep turns, chandelles, lazy eights, and spins, may be approved, provided the airplane is operated according to the applicable utility category limitations. These limits usually consist of a lower maximum takeoff weight and/or a smaller CG range. The aft CG limit usually is located forward of the limit approved for normal category operations, and other loading restrictions commonly apply to the utility category. For example, baggage or rear-seat passengers typically are not permitted.

When a normal/utility airplane is operated according to the limitations of the normal category, certain maneuvers often are prohibited. For instance, acrobatics, spins, or spins with flaps extended may not be authorized. All of these limitations are listed in the pilot's operating handbook, approved flight manual, and/or indicated by specific markings and placards. Since procedures vary with different airplanes, it is important to follow the specific recommendations of the manufacturer.

From this discussion, it is apparent that pilots need to know the specific limitations of their airplane for both normal category and utility category operations, when applicable. In an airplane that is certified for spins, a CG only slightly aft of the approved CG range could make recovery from a fully developed spin improbable. *If an airplane is not certified for spinning, there is no assurance that recovery from a spin is possible.*

FLIGHT REVIEW

The regulation establishing flight reviews was formulated in an attempt to maintain an acceptable level of pilot proficiency. The end result has been safer pilots and safer air travel. FAR Part 61.56 requires that each certificated pilot successfully complete a flight review consisting of a minimum of one hour ground instruction and one hour flight instruction every 24 calendar months. However, a person who has satisfactorily completed one or more phases of the FAA-sponsored Pilot Proficiency Award Program within the preceding 24 calendar months need not accomplish the flight review requirements. Flight instructors should encourage pilots to participate

in the FAA Pilot Proficiency Award Program which also is known as the Wings Program. It is described in AC 61-91G, *Pilot Proficiency Award Program*.

In addition, flight instructors and pilots who have completed a refresher course or certain pilot proficiency checks within the 24-month review period may be exempt from some or all of the flight review requirements. For example, a person who holds a current flight instructor certificate and who has completed an approved flight instructor refresher course within 24 calendar months, need not accomplish the one hour of ground instruction. Also, a person who satisfactorily completes a pilot proficiency check conducted by the FAA, an approved pilot check airman, or a U.S. Armed Force for a pilot certificate, rating, or operating privilege is not required to accomplish a separate flight review.

However, the FAA recommends that pilots consider also accomplishing a review under some of these circumstances. For example, a pilot with an airplane single-engine land rating may have recently obtained a glider rating, but may still wish to consider obtaining a flight review in a single-engine airplane if the appropriate 24-month period has nearly expired.

A flight review should actually begin in the initial contact phase. The instructor should interview the pilot by asking questions that cover three basic areas. First, the instructor should ask the pilot about the *type of equipment flown*. Some of the maneuvers and procedures reviewed in a single-engine airplane may be different than those in a light twin. The *nature of flight operations* should also be determined. For example, the areas of operation emphasized for a pilot who normally flies locally out of an uncontrolled airport should be different from those used for a pilot who primarily flies long cross-countries to busy airports located within Class B and Class C airspace. And finally, the instructor should ask about the *amount and recency of flight experience*. A private pilot who does not periodically practice stalls or short field landings may need more flight review time dedicated to these procedures.

A flight review should not be considered a testing session; rather, it should be a learning experience for the pilot. It is the pilot's opportunity to learn the recent changes in regulations and procedures and to update flying skills. Because this is not a test situation but an instructional service, the instructor should attempt to create a relaxed learning environment during both the preflight discussion and the actual flight.

To prepare the pilot for the flight review, the instructor should describe the method and sequence which will be used for conducting the review. Normally, a flight review should include the following steps:
1. Analysis of the pilot's general and recent flying experience
2. Assignments for academic study, including a review of FAR Part 91
3. Preflight oral discussion
4. Description of appropriate flight maneuvers and procedures
5. Performance of flight maneuvers and procedures
6. Postflight evaluation, recommendations, and logbook endorsement

Advanced preparation is very important for successful completion of the flight review. After analyzing the pilot's flight experience, the instructor should provide an assignment of appropriate areas for academic study. Instructors should refer to AC 61-98A — *Currency and Additional Qualification Requirements for Certificated Pilots*, for guidance. Among other recommendations, this AC contains a list of areas in FAR Part 91 which are pertinent to the review. The assignment also should include video presentations, films and/or chapters from private or instrument/commercial textbooks, appropriate FAA publications, and the pilot's operating handbook. In addition, a series of essay-type questions should be used during the preflight oral discussion.

The final step before beginning the flight portion of the review should be a discussion of the maneuvers and procedures which will be performed. Each maneuver should be explained, including the method of performance and completion standards. To help in the selection of the appropriate maneuvers and determination of completion standards, the Private or Commercial Pilot Practical Test Standards may be consulted. The test standards should not be used as structured outlines for the review, but selected items should be used to sample the pilot's performance. The instructor should review the pilot's logbook to determine total flight time as well as type and recency of experience in order to evaluate the need for particular maneuvers and procedures in the review. AC 61-98A provides a list of maneuvers for various categories and classes of aircraft which can be used as a guide in selecting areas considered critical to safe flight.

If the pilot does not display the necessary skill or understanding of a maneuver, time should be taken to explain, demonstrate, and practice the maneuver or procedure. In this manner, the flight review becomes a true learning experience and meets one of its main objectives.

The postflight evaluation should be thorough and objective. The instructor may include recommendations for resolving appropriate problem areas. Recommendations for further practice may be accomplished by the pilot during personal flying or, preferably, in subsequent training sessions with the instructor. In addition, the pilot should be encouraged to participate in periodic refresher training.

An endorsement must be placed in the pilot's logbook when an adequate level of knowledge, safety, and proficiency in the airplane has been demonstrated. the recommended endorsement is shown in Chapter 3, Section A — Authorized Instruction and Endorsements. If the review is not satisfactorily completed, the logbook entry should indicate the flight as a dual lesson. At no time should an entry indicate an unsatisfactory flight review.

SECTION B—AERODYNAMICS OF FLIGHT

The study of aerodynamics is a highly technical discipline which, traditionally, has been reserved for scientists and aeronautical engineers. However, aviation pioneers realized that pilots and, particularly, flight instructors required a specific working knowledge of aerodynamic principles. Aerodynamics is a valuable tool which the instructor should use throughout the training program. Many benefits will be derived if a working knowledge of the subject is imparted to the student. For example, an understanding of aerodynamics helps to shorten the training program and dispels many of the mysteries and fears associated with learning to fly. Most importantly, pilots who understand aerodynamic principles are more familiar with the performance capabilities of a particular airplane and, therefore, conduct flight operations with a higher degree of safety. The purpose of this section is to review and highlight the aerodynamic concepts and principles which are useful to the flight instructor who is conducting training in single-engine airplanes. Aerodynamic principles unique to multi-engine airplanes are described in Chapter 5.

BASIC FORCES

Most student pilots are intrigued by the relationship of the four forces acting on an airplane in flight. An understanding of the concept that thrust equals drag and lift equals gravity during straight-and-level, unaccelerated flight is the basic framework for building a working knowledge of aerodynamics. Any inequality between thrust and drag during straight-and-level flight results in acceleration or deceleration until equilibrium is attained. Climbs and descents are also the result of unequal thrust and drag. An airplane's climb performance and ceiling result from its capability of producing excess thrust in relation to drag.

During a constant rate climb or descent, aerodynamic forces are in equilibrium even though the altitude is changing. If this were not so, the rate of climb or descent would be accelerating or decelerating. This is a difficult concept to explain to students since they logically suppose the airplane "lifts" itself to cruising altitude. They also suppose that an airplane's service ceiling is determined by the ability of the wings to produce excess lift in relation to gravity. In the following discussion, we will examine each of the four forces in greater detail, discuss aerodynamic stability, and also analyze flight maneuvers in terms of aerodynamic principles.

LIFT

To aid in the discussion of lift, a review of basic airfoil terminology is presented in figure 1-18. Bernoulli's Principle states that, as the velocity of a fluid increases, the pressure within that fluid decreases. Since air must travel farther and move faster over the wing than under it, an area of low pressure is created above the wing. This contributes about 75% of the total lift. The impact pressure of the air that is deflected downward by the wing contributes the remaining lift. This results from Newton's Third Law of Motion which states that for every action there is an equal and opposite reaction. Figure 1-19 illustrates the forces which contribute to lift.

For a wing to generate lift, it must create a circulation in the airstream. This pattern is created, in part, by the shape and angle of attack of the wing. As the angle of attack increases, airflow velocity over the top of the wing increases and the pressure decreases, thereby producing the major portion of lift.

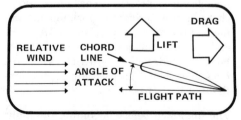

Fig. 1-18. Airfoil Terminology

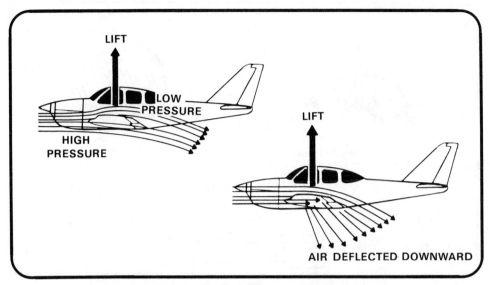

Fig. 1-19. Bernoulli's Principle and Newton's Third Law of Motion

AERODYNAMIC PITCHING MOMENTS

Under various conditions of lift production, such as flap extension, wings create pitching or twisting moments. The amount of these moments is controlled by the relationship between the aerodynamic center and center of pressure.

The *aerodynamic center* is the point on the wing chord where the aerodynamic forces balance during flight. All changes in lift effectively take place at this point. With a symmetrical wing, the upper and lower-surface lifts are opposite each other and pass through the aerodynamic center. With a cambered wing, the upper and lower-surface lifts are not opposite and cause a pitching moment about the aerodynamic center.

The point along the wing chord where the distributed lift is considered to be concentrated is called the *center of pressure*. When determining this point, all of the pressures above and below the wing are considered.

As the angle of attack changes under different flight conditions, the center of pressure also changes. The airfoil illustrated in figure 1-20 is an efficient cambered type used on many light general aviation aircraft. This figure shows that the center of pressure moves forward (toward the aerodynamic center) as the angle of attack is increased. This causes a reduction in the nose-down pitching moment.

When flaps are lowered, the center of pressure moves a considerable distance aft. Due to this change, flap extension often causes a noticeable nose-down pitching tendency. This tendency normally is most apparent in low-wing airplanes. With high-wing airplanes, flap extension causes a downward deflected airflow which pushes down on the horizontal tail surfaces. This deflected airflow may compensate for the aerodynamic pitch change and be great enough to cause a nose-up pitching force.

Angle of attack, airstream velocity, drag, wing area and camber, and air density are some of the factors involved in the generation of lift, and are items of immediate interest to the instructor and student. Most of the items, except air

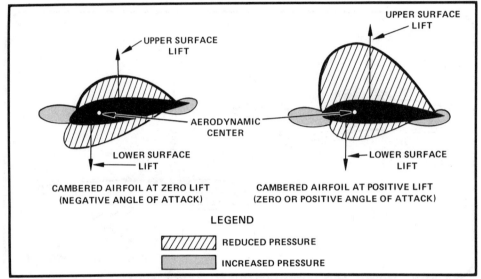

Fig. 1-20. Center of Pressure vs. Aerodynamic Center

density can be controlled, at least in part, by the pilot. Air density should always be considered in preflight planning because effects on aircraft performance are unique for each flight.

ANGLE OF ATTACK

Angle of attack is a prime consideration for all flight operations. As airspeed is decreased, the angle of attack must be increased to maintain the original lifting force. At some point, however, lift can no longer be increased in this way. Beyond this point, any further increase in angle of attack results in a separation of the airflow over the wing and a stall results, as shown in figure 1-21. Generally, a wing is designed so the stall occurs first at the wing root, then progresses outward along the span. This may be accomplished by the use of *washout* which provides a lower angle of incidence near the wingtips. Consequently, the wingtips have a lower angle of attack than the root and stall last. This provides adequate aileron control throughout the stall and during flight on the back side of the power curve.

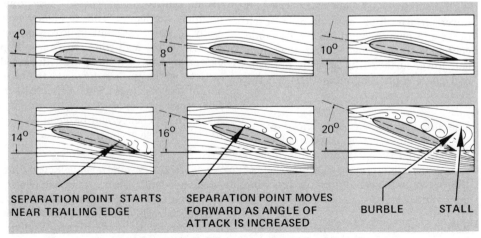

Fig. 1-21. Airflow Separation and Angle of Attack

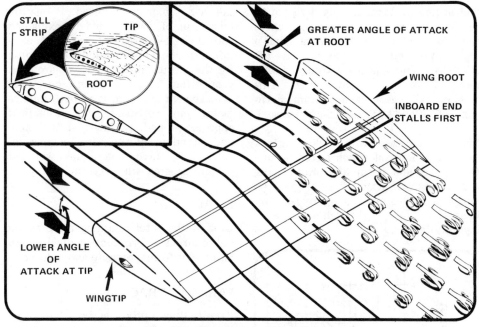

Fig. 1-22. Wing Design and Stall Patterns

Stall strips may be used to cause the same effect. As the angle of attack increases, stall strips break up the airflow and initiate the stall at the root. In other cases, a spanwise airfoil variation is employed which provides a high speed airfoil at the root and a low speed type at the tips. The low speed section can fly at higher angles of attack than the high speed section and, therefore, the root again stalls first. Figure 1-22 illustrates typical design methods used to make the wing roots stall before the tips.

WING PLANFORM

Planform is the shape of the wing as seen from above. Today's total airplane design is a series of compromises and wing design is no exception. Each wing shape has particular advantages and disadvantages, and usually is a combination of two or more basic shapes.

The efficiency of the wing is often described by the term *aspect ratio*. In general, the higher the aspect ratio, the more efficient the wing. It is computed by dividing the span by the average chord. The advantages and disadvantages of some basic wing planforms are noted in the following list.

1. *Straight wing*—excellent stall characteristics; most economical to build; inefficient from structural, weight, and drag standpoints
2. *Tapered wing*—efficient from structural, weight, and drag standpoints; stall characteristics are not as good as straight wing
3. *Elliptical wing*—most efficient from structural, weight, and drag standpoints; stall characteristics are not as good as straight wing; wing is more expensive to build than tapered wing
4. *Swept back and delta wings*—used for high speed aircraft that fly near the speed of sound; stall characteristics are more critical due to the high angles of attack required

The relative measure of an airfoil's lifting capabilities is termed the coefficient of lift (CL). It is the ratio between the airstream dynamic pressure and the lift pressure generated by the wing. This ratio is a function of the wing's design and angle of attack. A wing designed to produce lift at slow airspeeds, such as the type used on a STOL aircraft, has a high maximum coefficient of lift (CL_{MAX}). The opposite is true of a wing designed for high speed aircraft. This type of wing has a low maximum coefficient of lift and stalls at higher airspeeds.

Since lift increases with angle of attack up to the point at which a stall occurs, CL_{MAX} occurs at, or just prior to, the stalling angle of attack. The stall may be delayed, as illustrated in figure 1-23, by increasing CL_{MAX} through the use of high-lift devices.

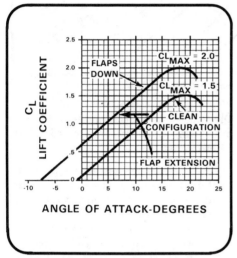

Fig. 1-23. Maximum Coefficient of Lift

HIGH-LIFT DEVICES

Although higher cruise speeds are the design goal of many airplanes, the high speed wing design does not lend itself to low approach and landing speeds. Therefore, to increase the safety and utilization of modern aircraft, various high-lift devices such as slots and trailing edge or leading edge flaps have been incorporated in their design. Trailing edge flaps are the most commonly encountered high-lift device in light general aviation type aircraft.

Each student must receive thorough training in the proper use of flaps and

the aerodynamics involved. Premature retraction of flaps should be explained and then demonstrated in flight at a safe altitude. This knowledge is essential for the execution of full flap go-arounds. Proper flap, airspeed, pitch, and power management also are vital performance factors in soft-field and short-field approaches and landings. A complete understanding of the aerodynamics involved in these techniques will aid student progress while insuring safety.

DRAG

Drag is the total retarding force acting on an airplane. Drag may be divided into two major forms: parasite and induced.

PARASITE DRAG

Parasite drag is produced by the shape of the aircraft (form drag), the surface of the aircraft (skin friction), and by interference of the airflow between parts of the aircraft (interference drag). Total parasite drag increases with airspeed.

INDUCED DRAG

Induced drag is created by the production of lift and increases as airspeed decreases, but at a much faster rate. It is inversely proportional to the wing span; for example, a long narrow wing produces less induced drag than a short wing. Many aircraft today are designed with tip tanks or end plates which provide the same effect as increasing the length of the wing; that is, they reduce induced drag.

LIFT-TO-DRAG RATIO

Minimum total drag occurs at the airspeed at which induced drag and parasite drag are equal. This is the point at which the airplane is producing the least drag for a given amount of lift and is said to be operating with the best lift to drag ratio (L/D_{MAX}). This airspeed also results in the maximum range, and is usually the best glide speed.

WINGTIP VORTICES

Wingtip vortices are caused by high pressure air below the wing flowing up and over the wingtip toward the low pressure air above the wing. Figure 1-24 demonstrates this principle. The tip vortices generated by large aircraft are often referred to as wake turbulence and are extremely dangerous to light aircraft. Wingtip vortices increase in velocity as speed decreases and load factor increases. From a performance viewpoint, vortices increase parasite drag and reduce wingtip efficiency.

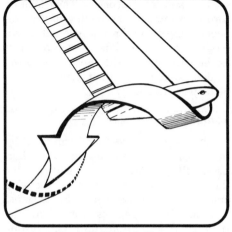

Fig. 1-24. Wingtip Vortex

GROUND EFFECT

Ground effect is encountered when operating at a height above ground within approximately one wing span of the aircraft. At these heights, the ground provides a restriction to the airflow around the wing, altering the flow pattern and wingtip vortices. As this occurs, the aerodynamic characteristics of the wing are changed. Induced drag is reduced and a smaller angle of attack is required for a given value of lift. The longitudinal stability of the aircraft also is affected by ground effect. For any given trim setting, the aircraft is more nose heavy while flying in ground effect.

The instructor must insure that each student thoroughly understands ground effect as it pertains to both takeoffs and landings. If students understand these principles, their takeoff and landing techniques will develop more rapidly.

THRUST

Thrust is the force that must be produced to overcome total drag. In a steady-state flight condition, thrust equals total drag. In the case of propeller-driven aircraft, thrust available depends directly on engine shaft horsepower and propeller efficiency. For an aircraft to climb, there must be an excess of thrust in relation to drag. The rate of climb depends on the amount of excess thrust available. Conversely, the rate of descent in a glide depends on a deficit of thrust in relation to drag. The more the power is reduced, the greater this deficit becomes, resulting in an increased rate of descent.

PROPELLER EFFICIENCY

Each blade of a propeller is essentially a rotating wing. As a result of their construction, the propeller blades produce forces that create thrust to pull or push the airplane through the air. The thrust which is created is a result of the propeller blades producing lift parallel to the airplane's longitudinal axis. The lifting action imparts a momentum change on the airstream. The highest propulsive efficiency is obtained when a large mass of air is moved with a relatively small velocity change.

If a propeller could perform with 100 percent efficiency, all of the engine's power output to the propeller would be converted to thrust. Therefore, it is important to maintain high propeller efficiency, as well as high horsepower.

Propeller efficiency is determined by numerous factors. Since the propeller is an airfoil, its *angle of attack* is the principal factor in determining efficiency. The angle of attack of a propeller is determined by the propeller pitch (called blade angle), propeller r.p.m., and the airflow along the propeller axis. These factors are illustrated in figure 1-25.

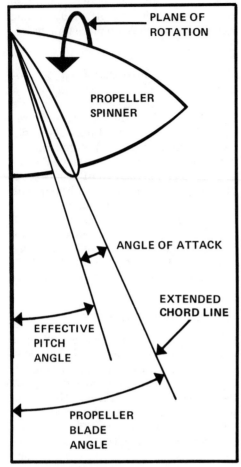

Fig. 1-25. Effective Pitch Angle

The variable-pitch propeller allows the *blade angle* to be changed, as necessary, to provide the desired angle of attack. However, since the angle of attack also is determined by the effective pitch angle, a change in either the r.p.m. or airspeed will cause a change in the angle of attack and, therefore, a change in propeller efficiency.

When an aircraft equipped with a fixed-pitch propeller accelerates, as in a dive, the propeller r.p.m. tends to increase. Additionally, if the airspeed decreases, the propeller r.p.m. tends to decrease. If the r.p.m. of a variable-pitch propeller is to be held constant, the propeller blade must be designed so that the *blade angle* may automatically be *increased* with an

increase in airspeed. Conversely, it must be designed so that the blade angle may automatically be decreased as the airspeed decreases. Maintaining a constant r.p.m. by variation of the propeller blade angle also tends to maintain the desired angle of attack and optimum propeller efficiency.

LEFT-TURNING TENDENCIES

There are several factors inherent in airplane design that result in left-turning tendencies. The forces involved normally are most pronounced during high power settings and low airspeeds. They include torque, asymmetrical loading of the propeller, slipstream rotation, and gyroscopic precession. The instructor must explain the cause of these forces to the student because of their effect on airplane performance.

Torque

Torque is a reactive force generated by the engine as it rotates the propeller. Essentially, it results in a tendency for the airplane to rotate about the engine crankshaft and propeller. Since most American-built engines rotate to the right as viewed from the cabin, torque reaction tends to roll the airplane to the left.

To counter this force, which is constantly exerting a rolling moment on the airplane in flight, it is common practice to rig a slightly higher angle of incidence in the left wing. Only at a specific power setting and airspeed can this higher angle of incidence, or "wash in," compensate for the torque originating in the engine by providing the exact amount of necessary force. To provide the maximum benefit to the pilot, the power and airspeed values for which the wings are rigged are those used for cruising flight in the airplane concerned.

Because the extra lift demanded from the left wing causes a slight increase in drag, the airplane tends to turn to the left unless it is restrained by some compensating force. To provide such a force, the leading edge of the vertical stabilizer or fin is usually offset to the left, giving the same effect as a slight application of right rudder. The amount of this offset is just sufficient to overcome the left-turning tendency at cruising speed. Consequently, the airplane turns left during slow speed flight and turns right during high speed flight unless corrective control pressures are applied.

Asymmetric Thrust

As long as the propeller plane of rotation is perpendicular to the relative air flow, the thrust it produces is uniform about the plane of rotation. However, if it is held at an angle to the direction in which it is moving through the air, as in a nose-high climb, significantly greater thrust is produced on the right side of its plane of rotation.

During climbs, asymmetric loading of the propeller, often called "P-factor," is caused by the higher angle of attack on the descending propeller blades. This creates a higher thrust force from the descending blades which are to the right side of the airplane's centerline. Since the propeller is turning in a clockwise direction, the greater thrust generated by the descending blades produces a pronounced left-turning tendency.

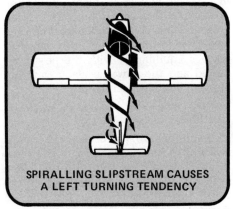

SPIRALLING SLIPSTREAM CAUSES
A LEFT TURNING TENDENCY

Fig. 1-26. Slipstream Effect

Slipstream Effects

The propeller slipstream is not a smooth, straight-line flow of air. It is a spiralling flow, turning in the same direction as the propeller. Because of this, the slipstream strikes the left side of the fuselage and vertical stabilizer causing the airplane to yaw to the left, as illustrated in figure 1-26.

Precession

The turning engine and propeller possess characteristics of a gyroscope. One of the properties of gyroscopic action is precession. *Gyroscopic precession* can be explained as the resultant action or deflection of the spinning object when a force is applied to this object. The reaction to a force applied to a gyro acts approximately 90° in the direction of rotation from the point where the force is applied. Therefore, if an airplane is rapidly moved from a nose-high pitch attitude to a nose-low pitch attitude, gyroscopic precession will create a tendency for the nose to yaw to the left. The left-turning force is quite apparent in tail-wheel type airplanes as the tail is raised on the takeoff roll. This is depicted in figure 1-27.

WEIGHT

Weight, or gravity, is the simplest of the four forces. This is the actual weight of the airplane, termed one-G, and it always acts downward, toward the center of the earth. The effective weight, or load factor, can be increased above the normal one-G by flight maneuvers or turbulence. For this reason, airplane structures are designed to withstand load factors much greater than the one-G weight of the airplane.

FORCES IN FLIGHT
STRAIGHT-AND-LEVEL FLIGHT

When an airplane is in straight-and-level, unaccelerated flight, a condition of equilibrium exists. Lift equals weight, and thrust equals total drag. To accelerate or

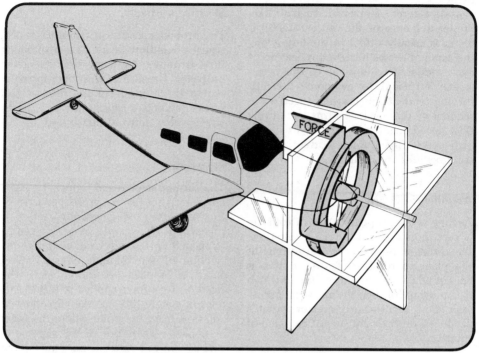

Fig. 1-27. Gyroscopic Precession

decelerate in level flight, the power must be changed to produce an unbalanced condition. But, as the power and airspeed change, an adjustment of the angle of attack must be simultaneously accomplished, or level flight will not be maintained.

Increases in airspeed require increases in engine power. Figure 1-28 shows the power required to achieve equilibrium at various airspeeds. If an airspeed equal to

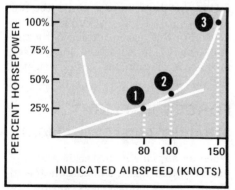

Fig. 1-28. Power Required vs. Airspeed

point 1 is desired, the power required curve shows the engine power that must be available to maintain equilibirum. If a speed equal to point 2 is desired, the power required for equilibirum is considerably greater. The maximum level flight speed, shown in point 3, is obtained when the power required equals the power available from the powerplant.

The airplane will continue to accelerate in straight-and-level flight until the power required equals the power available. At this point, the increase in drag associated with the increased airspeed reaches equilibirum with the power required and the power available, and the airplane no longer accelerates.

CLIMBS

Maximum climb performance is *not* a result of excess lift, but a result of excess thrust horsepower and/or thrust. The angle of climb is a function of thrust, while rate of climb is a function of thrust horsepower. During most phases of flight, the maximum rate of climb is used. As illustrated in figure 1-29 the maximum rate of climb occurs at the airspeed with the greatest difference between thrust horsepower available (product of shaft horsepower and propeller efficiency) and thrust horsepower required.

For any given airplane weight, the maximum angle of climb is dependent upon

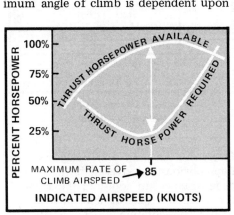

Fig. 1-29. Maximum Rate of Climb

the amount of excess *thrust* available. This occurs where there is the greatest differential between thrust and drag. Figure 1-30 shows the thrust available and required curves in relation to airspeed. It should be noted that thrust available decreases as the airspeed increases, but the thrust required curve increases with airspeed. Therefore, the maximum angle-of-climb airspeed occurs at the point with the greatest differential between thrust available and thrust required.

The thrust available curve shows that propeller thrust is the highest at low airspeeds and decreases as airspeed increases. The maximum excess thrust is, therefore, available at an airspeed very close to the stall speed.

FACTORS AFFECTING CLIMB PERFORMANCE

Airspeed affects both the angle and rate of climb. At a given airplane weight, a specific airspeed is required for maxi-

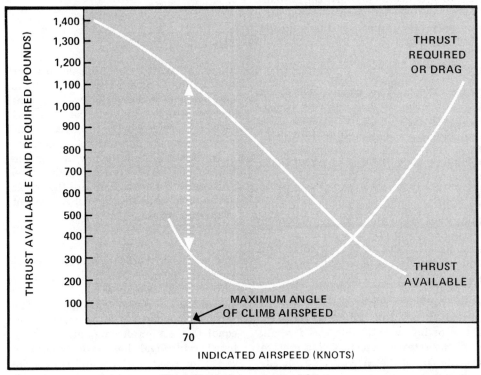

Fig. 1-30. Maximum Angle of Climb

mum performance. Generally, the larger the speed variation from that required, the larger the performance variation.

Weight also affects climb performance. A change in weight changes the drag and power required. Increases in weight reduce the maximum rate and angle of climb.

Increases in altitude generally have the greatest effect on climb performance. This occurs because an increased altitude increases the power and thrust required, but decreases the power and thrust available. Therefore, both the angle-of-climb and rate-of-climb performance decrease with altitude. At the same time, the airspeed necessary to obtain the maximum angle of climb increases with altitude, but the airspeed for best rate of climb decreases, as shown in figure 1-31.

The point where these two speeds converge is the airplane's *absolute* ceiling; neither excess power nor thrust is available to produce a climb. At the absolute ceiling, the rate of climb becomes zero, and only one airspeed results in steady, level flight.

GLIDES

In the event of an engine failure, the pilot of a single-engine airplane usually is interested in flying the airplane at the minimum glide angle to obtain the maximum distance for the altitude lost. This minimum glide angle is obtained under the aerodynamic conditions which produce the least drag. Since the lift is basically equal to the weight, the minimum drag is obtained at the maximum lift-to-drag ratio (L/D_{MAX}). This occurs at a specific value of the lift coefficient and, therefore, at a specific angle of attack.

Power-off glide performance is directly related to airspeed. Flight at airspeeds above or below best glide speed results in higher rates of descent and, therefore, less horizontal distance covered for altitude lost.

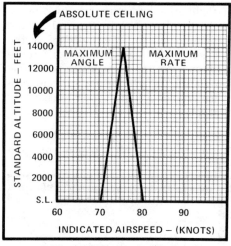

Fig. 1-31. Change in Climb
Speed with Altitude

Increasing drag by extending gear or flaps obviously decreases the maximum glide distance. In addition, the airspeed for maximum glide distance in that particular configuration is less, which further decreases the horizontal distance of the glide.

TURNS

Turns are made by banking the airplane to shift the vertical component of lift to one side and produce a force which turns the aircraft. The lift force can be subdivided and represented as two forces — one acting vertically and one acting horizontally. Figure 1-32 shows how the horizontal component of lift (centripetal force) accelerates the aircraft toward the center of the turn, while the vertical component overcomes weight.

In a level turn, there are pairs of opposing forces in balance. The centrifugal force acting on the aircraft is equal and opposite to the centripetal turning force. The vertical component of lift is opposite and equal to the aircraft's weight. The combined centrifugal force and weight is called the resultant force and is opposite and equal to the total lift.

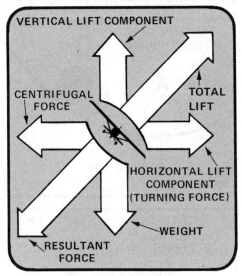

Fig. 1-32. Balanced Forces
in a Coordinated Turn

The steeper the bank, the greater the total lift needed and the greater the force required to make the airplane turn. This also means that the centrifugal and gravitational forces combine to make a greater resultant force and higher load factor, as illustrated in figure 1-33.

In level flight, the total lift force is equal to weight. When the aircraft is banked, the total lift is diverted from vertical. Since the total lift is still equal to weight, insufficient lift is acting vertically to counteract weight, and the aircraft descends. In order to maintain altitude in a turn, the total lift must be increased until sufficient lift acts vertically to counteract weight. This is done by increasing the angle of attack and, therefore, the coefficient of lift.

If the aerodynamic principles involved in turns are clearly understood, the student will have less difficulty learning the correct techniques of the turn. Such knowledge also will enable the student to better understand the function of angle of bank and its effect on stall speed.

AIRCRAFT AXES

The movement and control of the airplane in relation to its axes is a very basic but necessary orientation for the student pilot. Beginning pilots must understand that the three axes of an aircraft are identified by their alignment with the airplane and by the motion which results from the airplane's movement about them. For example, the axis line which runs from the nose to the tail of the airplane is called the longitudinal axis. Motion about that axis is called *roll* and is controlled by aileron displacement.

The axis from wingtip to wingtip is the lateral axis. Movement about this axis is *pitch*, which is controlled by the elevator. The vertical, or *yaw*, axis is defined by a vertical line through the intersection of the longitudinal and lateral axes. Yawing motion is the result of improper rudder displacement. To maintain coordination, the rudder normally is used in conjunction with the ailerons. It should be noted that all aircraft move-

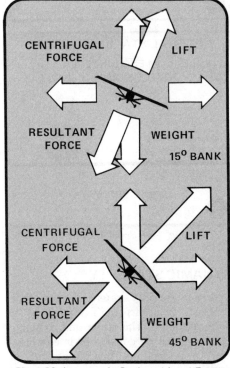

Fig. 1-33. Increases in Bank and Load Factor

ment in flight occurs about the center of gravity; therefore, the aircraft axes intersect at the center of gravity.

Preflight briefing and in-flight demonstrations are of equal importance in insuring that the student understands the functions of the control surfaces. One of the first flight lessons should include a demonstration of the response of the aircraft to the individual control surfaces. If students are able to visualize the relationship between control surface movements and airplane movement, they will progress more rapidly and become more proficient pilots. Many times a student's problems with advanced maneuvers are a result of a lack of understanding of these basic principles.

AIRCRAFT STABILITY

An airplane in flight is constantly affected by minor control pressures and outside forces that move it from the desired flight attitude. The tendencies of an airplane, once it is disturbed from an attitude, depend upon the type of stability that is designed into the airplane.

STATIC STABILITY

Static stability is the tendency of an airplane to return to a state of equilibrium following a displacement from that condition. If *positive static* stability is present, the airplane has a tendency to return to the original point of equilibrium. As shown in figure 1-34, part 1, when the ball is moved from point A, it has a tendency to return to the original position.

Part 2 illustrates *negative static* stability. When the ball is displaced from point A, it has a tendency to move farther from the point of equilibrium.

Neutral static stability is illustrated by part 3. When the ball is displaced from equilibrium at point A, it does not have a tendency to return to, or move farther from, its original point of equilibrium.

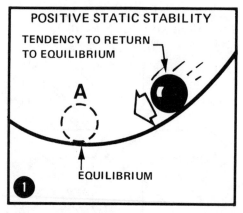

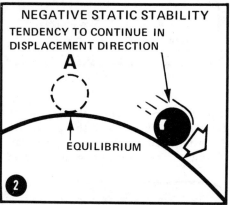

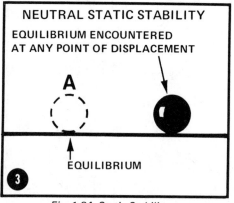

Fig. 1-34. Static Stability

DYNAMIC STABILITY

Dynamic stability refers to the time required for an airplane's response to its static stability. The static and dynamic stability must be considered in combination to understand the effects on an airplane.

Airplanes are designed with *positive static* and *positive dynamic* stability. This results in the airplane's tendency to return to equilibrium through a series of decreasing oscillations, as shown in figure 1-35, airplane A. If airplane B displays positive static, but *neutral dynamic* stability, it attempts to return to equilibrium but the oscillations do not decrease in magnitude with time; therefore, equilibrium is not regained. Positive static and *negative dynamic* stability are indicated by airplane C. After the displacement from equilibrium occurs, the oscillations increase in magnitude with time.

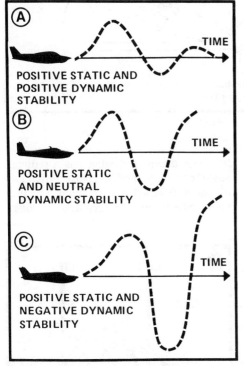

POSITIVE STATIC AND
POSITIVE DYNAMIC
STABILITY

POSITIVE STATIC
AND NEUTRAL
DYNAMIC STABILITY

POSITIVE STATIC AND
NEGATIVE DYNAMIC
STABILITY

Fig. 1-35. Oscillatory Stability

LONGITUDINAL STABILITY

If the aircraft is statically stable along its longitudinal axis, it will resist any force which might cause it to pitch, and it will return to level flight when the force is removed. This is the most important of the three types of stability of an aircraft in flight.

To obtain longitudinal stability, an airplane may be designed to be either slightly tail-heavy or nose-heavy while trimmed in straight-and-level flight. Additionally, the CG location can affect the nose-heavy or tail-heavy condition. If the airplane is loaded near the forward CG limit, a nose-heavy condition can occur. Conversely, if it is loaded near the aft CG limit, a tail-heavy condition may result. Consequently, the horizontal stabilizer is designed with the capability of producing either the positive or negative lift necessary to longitudinally stabilize the airplane within CG limits.

If an airplane is tail-heavy during straight-and-level flight, the horizontal stabilizer produces positive lift to maintain longitudinal stability. If turbulence or control input pitches the nose up, the horizontal stabilizer experiences an increased angle of attack, as shown in figure 1-36, part A. This results in an increase in the horizontal stabilizer's positive lift, forcing the nose down.

If the airplane is nose-heavy, the horizontal stabilizer produces negative lift to maintain longitudinal stability. In this situation, a nose-up displacement also results in a lift change on the horizontal stabilizer. The new angle of attack of the stabilizer can result in a conversion from negative to positive lift or a reduction in negative lift, as illustrated in figure 1-36, part B. Either situation results in raising the airplane's tail and lowering the nose. The opposite reaction takes place if the nose of the airplane pitches downward.

A horizontal stabilizer producing positive lift experiences a decrease in angle of attack; therefore, less tail lift is developed and the tail lowers as the nose raises. If the stabilizer is producing negative lift to maintain stability, the pitch change results in an increase in the angle of attack, causing the stabilizer to produce greater negative lift. This situation also forces the tail down, returning the airplane to the trimmed condition.

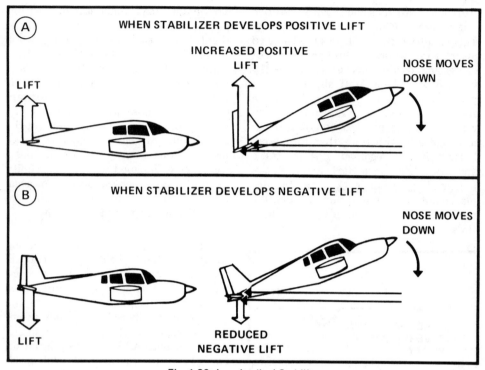

Fig. 1-36. Longitudinal Stability

LATERAL STABILITY

Lateral stability refers to an airplane's tendency to return to wings-level flight following a displacement. It is considered generally undesirable to design an airplane with *strong* lateral stability. Such stability has a detrimental effect on rolling performance and handling characteristics during crosswind take-offs and landings.

The tendency of an airplane to right itself after being displaced from wings-level flight actually is the result of a side slip which is induced by corrective control movements. If a wing is displaced, another force, such as a control input, must be introduced before the airplane's lateral stability becomes evident. As illustrated in figure 1-37, position 1, a wing develops stable rolling moments when a side slip is introduced. When the relative wind strikes the airplane from the side, the upwind wing experiences an increased angle of attack and increased

lift. The downwind wing has a reduced angle of attack and decreased lift. The differential in lift results in a rolling moment, which tends to raise the low wing.

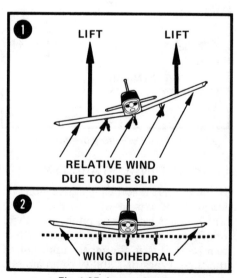

Fig. 1-37. Lateral Stability

Dihedral, as illustrated in figure 1-37, position 2, is a wing design consideration used to obtain lateral stability. Dihedral increases the stabilizing effect of side slips by increasing the lift differential between the high and low wing during the slip.

DIRECTIONAL STABILITY

Directional stability is the tendency of the airplane to remain stationary about the vertical or yaw axis. Figure 1-38, aircraft A, illustrates that when the relative wind is parallel to the longitudinal axis, the airplane is in equilibrium. If some force yaws the airplane, producing a side slip, a positive yawing moment also is developed, which returns the airplane to equilibrium.

The vertical stabilizer is a symmetrical airfoil, capable of producing lift in either direction. As shown by aircraft B, a positive side slip angle changes the relative wind on the vertical stabilizer, creating a higher coefficient of lift in one direction. In the illustration, the airplane is in a side slip to the right, resulting in positive lift to the left of the stabilizer. Additional lift moves the tail to the left, causing the longitudinal axis to align with the relative wind.

To obtain this stability, the side area of fuselage ahead of the center of gravity must be less than that behind the center of gravity. This difference in fuselage area is usually insufficient for a high degree of stability, so a vertical stabilizer is added.

This section has attempted to present the broad range of aerodynamic principles which are useful to instructors training in single-engine airplanes. It is not

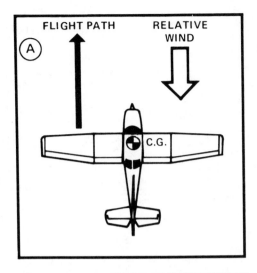

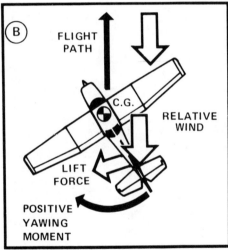

Fig. 1-38. Yaw and Side Slip

intended to be an exhaustive treatment of the subject. However, the material should provide the necessary framework for beginning instructors to increase their personal knowledge of aerodynamics and to apply aerodynamic principles to the process of flight instruction.

chapter 2

THEORY OF INSTRUCTION

INTRODUCTION

The necessary qualifications for a flight instructor go far beyond the ability to competently pilot an airplane. The professional flight instructor must acquire a working knowledge of how people learn, anticipate their behavior patterns, select the best methods for guiding them through the learning process, and develop the ability to judge their performance in ways that enhance learning. A careful study of the principles, techniques, and methods summarized in this chapter will help instructors understand their students, improve the quality of their instruction, and enhance their ability to communicate. This chapter highlights the principles of learning and teaching best suited to flight and ground instruction and also outlines specific ways to plan and organize learning experiences from single lessons to entire courses.

SECTION A — THE LEARNING PROCESS

It is a fundamental assumption that a cross-country flight must have a destination. The destination is the purpose of the trip, the goal or, more specifically, the objective. It is obvious that preflight planning and other preparations cannot proceed until the destination (objective) has been defined. In a similar manner, an objective must be formulated before starting on a training flight or a period of ground instruction. The objective should contain sufficient information so a competent person can correctly interpret the purpose of a given period of instruction.

OBJECTIVE

An objective can be defined as the *destination* of a specified amount of activity. For a trip, it is stated as an airport or geographical coordinates. In either case, its designation leaves no room for doubt in the mind of a competent person. An instructional objective must be equally specific. If written in vague terms, there can be as many interpretations as there are people available to interpret it. Specific objectives must state *the intended change in the student's behavior*. When defined in terms of what the students

- STATES WHAT THE STUDENT WILL DO TO PROVE THAT THE OBJECTIVE HAS BEEN ATTAINED
- STATES THE CONDITIONS UNDER WHICH IT WILL BE ATTAINED
- STATES MINIMUM ACCEPTABLE PERFORMANCE

Fig. 2-1. Qualities of Useful Objectives

are able to do on completion of the instruction, an objective enables the instructor to evaluate progress. When this objective is communicated to the students, they, too, can determine progress.

QUALITIES OF USEFUL OBJECTIVES

A specific objective must describe the intended change in the student's behavior. In order for the instructor to evaluate this change, it is necessary for the learner to *do* something. Such words as "know," "appreciate," and "understand" do not, in themselves, express an observable result. One of the most frequently used words in stating objectives is "demonstrate." This may be modified by "orally," or "in flight," to describe the conditions under which the student may be asked to demonstrate, thus specifying the action to be performed. Other useful words are "recite," "describe," "select," and, in the case of flight maneuvers, "perform." A sample objective for a

period of instruction in aerodynamics, then, could include the statement, "The student will list the four forces acting on an airplane in flight." To "understand" the forces would not constitute an observable change.

A useful objective also contains a description of the conditions under which the intended behavior is to occur, such as, "S-turns across a road in a 20-knot wind." And, the observable behavior should be *measurable*. For instance, an objective for a course of ground instruction might state, in part, "Upon completion of this period of instruction, the student will be able to answer correctly 75 to 80 questions on FAR Part 61 in a period of two hours, without the use of reference material." To verify attainment of this type of objective, some instructors give sample questions during the course introduction to show the students exactly what is expected. A summary of the qualities of a useful objective is shown in figure 2-1.

Major objectives for the outcome of a total course are contained in the appropriate written test and Practical Test Standards. What a student is expected to do, the conditions under which it will be done, and minimum performance standards are outlined clearly. A training syllabus oriented toward those objectives will guide the instructor and the student toward the successful accomplishment of the requirements. Currently developed training syllabus usually specify the acceptable performance criteria in the lesson completion standard. The completion standard relates directly to and is considered part of the objective.

In order to fulfill the training syllabus requirements most efficiently, however, the instructor must establish intermediate goals for each period of flight and ground instruction. A flight instructor applicant is required to complete a lesson plan for a period of flight instruction specified by the examiner. The first item on the lesson plan is the goal, or objective, of that period of instruction. A well-written objective, containing all the elements discussed, makes the rest of the plan easy to execute. When the goal is described clearly, the equipment and procedures used become obvious and the evaluation is already written.

A session conducted according to the lesson plan is the one which can be expected to yield the best results. Most pilots have not-too-fond memories of an instructor who rushes up to them 5 or 10 minutes late exclaiming, "Hi, sorry I'm late. Listen, why don't you run out and do a preflight and I'll be with you in about 10 minutes, and we'll go out and do some stalls." The confused student walks out to the airplane not knowing what to expect—envisioning the airplane falling out of the sky. It is much better for all concerned when the instructor is prepared to sit down with the student in a brief preflight session which explains what will be expected during the flight.

ATTAINING THE OBJECTIVE

Attaining the objective requires selection of the methods and techniques to be used in the teaching process. The methods an instructor should be familiar with are based on many areas of knowledge, including the theory of learning, human behavior, effective communications, and specific techniques which have worked successfully for others.

LEARNING THEORY

The ability to learn is an extraordinary human characteristic. Learning occurs continuously throughout a person's lifetime and as a result of each learning experience an individual's way of perceiving, thinking, feeling, and doing may change. Thus, learning can be defined as a change in behavior as a result of experience.

LEARNING CHARACTERISTICS

Students will learn from any activity that tends to enhance their own purposes. Each student is an individual, with a unique composition of background, goals, ambitions, and motivations. Every student has a different reason for learning something and each anticipates different end results. The effective instructor is the one who finds ways to relate the instruction to the particular needs of the student. If a student is just sliding by, chances are that no connection has been made between the purpose of the subject matter and the student's desires.

Learning comes through experience; it is the result of an individual's experiences. It cannot be forced by someone else. Knowledge is a result of experience, and no two people have the same experience, even when watching the same event. Individuals learn from an experience according to the way it affects their individual needs and the way past experience has conditioned them. The degree of learning resulting from an experience depends, to a great extent, on the depth of the experience. Most people, for example, have difficulty memorizing lists of words. However, an idea, or concept which strikes a person as being useful will be easily assimilated. An experience which challenges and involves the student's thoughts, feelings, past experiences, and physical activity is far more effective than one which requires only that the student memorize statistics. With this in mind, the instructor must show the student the relationships between various subject areas and personal safety or enjoyment during flight.

A single learning experience may include verbal elements, as well as perceptual and emotional elements. When learning to do stalls, for example, the student's understanding of angle of attack discussions in the classroom will be augmented. While trying to solve a particular problem, many people have suddenly gained insight into something related to an entirely different situation. Thus, the instructor who sees the objective only as training for a particular skill or area of knowledge may be missing some of the potential opportunities for the training session.

Learning is an active process. The instructor who expects a student to sit idly in class and soak up knowledge like a sponge should not be dismayed to discover that the knowledge, like the water in the sponge, evaporates overnight. If the learner is to retain the knowledge, there must be active involvement in some manner. Learning has been defined as a change in behavior. Certainly, then, the process must be an active one.

LAWS OF LEARNING

There are numerous laws of learning. They are included here to assist the instructor in gaining insight into how and why people learn.

LAW OF READINESS

The law of readiness states that *a person learns when ready to learn, and a person is ready to learn when a reason for learning is perceived.* Some students may be ready to learn even before they walk into the classroom, while others may be interested, but not ready to learn. The instructor often must instill and nurture the desire to learn.

There are circumstances, of course, under which even the most strongly motivated person is not ready to learn, and there is little, if anything, that the instructor can do about it. If outside influences, such as health, finances, or family problems weigh heavily on the student, the desire to learn can be completely overshadowed.

LAW OF EXERCISE

The law of exercise states that *those things most often repeated are best remembered.* Thus, practice and drill are absolute requirements for learning. The instructor, then, must provide opportunities for practice, insuring that it is directed toward a goal. Types of practice may include recall, review, restatement, manual drill, and physical application.

LAW OF EFFECT

The law of effect is related to the emotional reaction of the student. According to this law, *learning is strengthened by a pleasant or satisfying feeling and weakened if associated with an unpleasant feeling.* Setting goals that students cannot reach, or introducing material beyond comprehension, will result in feelings of frustration, futility, and inadequacy.

Instead, the instructor should establish intermediate goals the students are capable of reaching, so they can experience success and feel they are making progress toward the ultimate goal. Negative motivation should not be used, because fear of an undesired outcome does not strengthen learning. It is the anticipation of a successful outcome that prompts students to do their best. Although every learning experience may not be successful, students should not be allowed to feel degraded by a failure.

LAW OF PRIMACY

According to the law of primacy, *that which is learned first often creates an almost unshakable impression.* A person with the impression that a particular thing is no good, whether the opinion is well founded or not, will be much more difficult to convince that it really works than is one who had not formed an opinion. A student with an improper landing technique is more difficult to teach than the one who has

formed no specific ideas about proper landings. Due to this factor, the instructor must teach the student properly the first time, so that at some later date, the double task of unlearning the wrong and then learning the correct method will not be necessary.

LAW OF INTENSITY

The law of intensity states that *an exciting experience teaches more than a boring experience.* The more a learning experience is related to the real thing, the more valuable it is, because it is vivid and stimulating. It is easy to put a student to sleep while talking about flying, but attention rarely lapses while practicing precision maneuvers. It is up to the classroom instructor to use imagination and innovations such as visual aids and panel discussions.

LAW OF RECENCY

The law of recency indicates that *when other things are equal, that which was learned most recently is the easiest to*

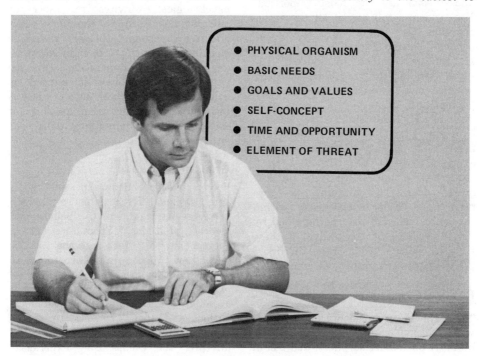

- PHYSICAL ORGANISM
- BASIC NEEDS
- GOALS AND VALUES
- SELF-CONCEPT
- TIME AND OPPORTUNITY
- ELEMENT OF THREAT

Fig. 2-2. Factors Which Affect Perceptions

recall. A fact learned but not recalled or used in a particular application for an extended period of time is difficult to recall. A telephone number dialed moments before is easy to recall, but two hours later may not be recalled at all. The instructor applies the law of recency when summarizing the important points of a lecture to assist students in remembering them.

HOW DOES LEARNING OCCUR

Learning has been defined as a change in behavior as a result of experiences. To understand how experiences change behavior, it is helpful to know some of the factors involved in the learning experiences and how they affect the learning process.

PERCEPTION

Perception is the basis of all learning. Perceptions may be directed to the brain by one or more of the senses. In flight training, the eyes, ears, and the muscular sense of changing position (kinesthesia) are the most important. The student's full use of all perceptive channels is important, since the behavior change sought is the result of all the perceptions introduced. The sense inputs to the brain initially produce only awareness; they become perceptions when they are given meaning. For example, a slow decrease in engine r.p.m. on a cold, humid day means something different to an experienced pilot than it does to a student who has flown only on less humid days. The meaning of these perceptions, then, comes from within the individual, and may be modified by several factors. A knowledge of these factors, which are illustrated in figure 2-2, is very important to an instructor.

Physical Organism

The physical organism is the means by which awareness is conducted to the brain. A pilot must see, hear, and feel in order to respond to existing circumstances. If any of that capability is impaired, some responses may be inaccurate. Flight physicals are aimed at determining the efficiency of the perceptual apparatus.

Basic Needs

A person's basic need is to maintain and improve both the physical and psychological portions of the total self. The food eaten and the air breathed become part of the physical self. In like manner, the sensory inputs become part of the psychological self. People develop reactions to prevent those things which would damage them physically, such as warding off a blow. Similarly, they develop perceptual barriers to ward off sights, sounds, and other inputs which could damage them psychologically.

To help a student learn, it is necessary to find ways to develop the desired perceptions, despite the defense mechanisms. Asking students to do something they believe will imperil them either physically or psychologically invites resistance or outright denial. It is, therefore, more effective to attempt to work with this basic need than to try to fight it. As an example, less cooperation can be expected if stalls are introduced as being dangerous than if their practice is introduced as a means of avoiding or escaping from a potentially dangerous situation.

Goals and Values

Perceptions depend on a person's goals and values. Everything a person perceives is modified by beliefs and mental attitudes. Spectators at a sporting event, for instance, notice infractions of the rules according to which side they are supporting. Because of this, it is important for instructors to attempt to determine the philosophical outlooks of their students to enable them to predict how the students will interpret the events they experience. Motivation also is affected by the person's goals and values. Since motivation is one of the most important factors in learning and is affected by many other factors, it will be discussed in detail in another section of this chapter.

Self-Concept

Self-concept, or the way people perceive themselves, is one of the most powerful factors affecting learning. It is closely related to the psychological aspects of basic needs. If students believe that an activity will perpetuate and improve their image of themselves, they enter into it wholeheartedly. When students feel insecure, their confidence must be bolstered by repeated successful experiences before attempting to perform a difficult task. Confident persons, on the other hand, are willing to try anything and, if they fail, are willing to try again, with no sense of damage to their self-concept. Students who view themselves positively are less defensive and more ready to assimilate all aspects of the instruction and demonstration presented.

Time and Opportunity

Time and opportunity are required for perception. To a great extent, learning depends on perceptions which have preceded the current effort. Sequence and time are necessary to enable the student to relate current perceptions to those which preceded them. If each flight maneuver is built on one or more which preceded it, both time and sequence are provided as the student practices the new maneuver. In the event of difficulty, lengthening the practice time might appear to be the obvious solution. This is true only if improvement continues to be evident, and the student's perceptions are not dulled by fatigue. The efficiency of a properly planned syllabus is proportional to the consideration given to the time and opportunity factor.

The Element of Threat

Threat restricts perception because fear not only affects responses, but also narrows a person's perceptual field. Fear of the unknown results of a stall might cause a student to stare at the airspeed indicator and overreact to decreasing airspeed indications. In so doing, a student might neglect a heading correction and fail to check for other traffic or obstructions. Any action by the instructor which the student sees as threatening reduces the ability to accept the experience. All of the student's physical, emotional, and mental faculties are adversely affected. The old method of trying to frighten a student into better performance has been proven to be psychologically wrong. The effective instructor organizes the logic of the instruction to accommodate the psychology of the student. If the student feels overwhelmed by the situation, a threat appears to exist and the capacity to learn is restricted. When the student feels capable of coping with the situation, the experience is considered a challenge, and there is motivation to attempt it.

Behavior is affected by the individual's perceptions which, in turn, are affected by any or all of the foregoing factors. A good instructor facilitates the learning process by avoiding those actions which may seem to threaten the student. Teaching is most consistently effective and efficient when all of the factors affecting perception are recognized and taken into account.

INSIGHT

Insights are the result of combining related perceptions into meaningful units. The perception is the "what," and the insights provide the "how,"and "why." Obviously, if the "what" is incorrect, the other two factors are also in error, and an erroneous insight is developed. It is the instructor's prime responsibility to develop correct insights and to build one upon the other until the goal is achieved.

The student must be kept constantly aware of insight-building perceptions and their relationship to the other significant pieces of the total picture. As an example, suppose the instructor is attempting to reach the concept "attitude + power = performance." From straight flight at cruising airspeed, the student is supposed to pull the nose up slightly and observe the result. The student perceives that airspeed decreases and the airplane begins to climb. (With fixed-pitch propeller, the r.p.m. also decreases.) If the nose is raised further, airspeed continues to decrease, but a point is reached where the airplane no longer climbs and begins to "mush." The student perceives that the nose of the airplane cannot be raised too much without suffering a loss of performance. Adding power and holding the attitude constant, however, increases both airspeed and rate of climb. Putting all of these perceptions together, insight is developed into the objective concept that "attitude + power = performance." This can be further developed into the more operational concept that to achieve any desired performance requires coordination of pitch attitude and power.

By building one perception onto another, the instructor successfully develops an understanding of the effect of pitch attitude and power on airspeed and rate of climb, and the student learns what action is necessary to achieve the desired performance. As a student's understanding develops, some insights inevitably are gained without assistance. The purpose of instruction, however, is to speed the process by pointing out the relationships of the perceptions as they occur and by promoting the development of insights. As useful perceptions increase in number and are assembled into insights, the rate of learning increases.

Forgetting is less of a problem as the student builds a background to which new experiences can be related. Pointing out the relationships as they occur, providing a challenging, but not threatening environment, and helping to build a favorable self-image are all vital in the process of helping the student develop insights.

MOTIVATION

Motivation causes people to set goals, then pursue them. Without goals, people stagnate; they have no place to go. Highly motivated persons, however, know where they want to go and how they plan to get there, and they constantly strive toward their goal. They meet the obstacles in their path and overcome them. Motivation is the dominant force governing a student's progress and ability to learn. Motivations occur in degrees, from zero to such power that they dominate every waking moment. However, motivation may also be negative, engendering doubts, fears, and defense mechanisms.

One of the tasks of the instructor, then, is to provide positive motivations. Reward of some sort is a means of motivating a person, but the type of reward depends on the individual's own values. While some are motivated strongly by monetary gain, others may seek acceptance and approval by their peers. Still others may wish only to acquire an additional skill or talent. Whatever the reward, students must believe that it is worth the time and effort required to attain it, and it must be constantly apparent to them that it is within their reach.

Motivation is one of the reasons for communicating objectives to students. If they believe the activity in which they are engaged is not relevant to their goals, motivation suffers and the lack of enthusiasm for the job at hand reduces performance. If the objective and its significance in the pursuit of their goal are made clear to them, however, motivation remains high.

Everyone seeks approval, and positive reinforcement is one method of showing this approval. Something good can be found in almost any performance and it should be pointed out to the student. This does not mean that low performance levels must be accepted or that standards have to be lowered to accommodate a slow student. Errors may be pointed out privately in a courteous fashion, without ridicule. On the other hand, sincere praise given in the hearing of others is a strong motivating force. In any group activity, praising the good student in front of the rest of the group not only motivates the good student, but stimulates the others to strive for praise. Conversely, ridiculing a student for a wrong answer to a question can cause the rest of the class to withdraw to avoid the threat to their own dignity.

Positive motivation is essential to the accomplishment of learning. Negative motivation, such as threats and reproof, should be avoided with all but the arrogant or impulsive student. At times, in spite of the instructor's efforts, the learning rate may be retarded by a slump in motivation. When this occurs, instructors should examine their own conduct and try to find the cause for the slump. If the slump is caused by outside factors, instructors can only redouble their efforts at positive motivation in an attempt to overcome the outside factors.

LEVELS OF LEARNING

Learning may be accomplished at any of several levels. The lowest level, *rote learning*, is the ability to repeat back something which one has been taught, without understanding or being able to apply what has been learned. Progressively higher levels of learning are *understanding* what has been taught, achieving the *skill to apply* what has been learned and to perform correctly, and associating and *correlating* what has

been learned with other things previously learned or subsequently encountered. Correlation is the highest level of learning, and should be the objective of all instruction. It is the level at which the student becomes able to associate an element which has been learned with other segments or "blocks" of learning or accomplishment. The other segments may be items or skills previously learned, or new learning tasks to be undertaken in the future.

LEARNING A SKILL

Although skills are often classified as either mental or motor (physical) skills, each involves at least some of the other. Reading, for example, is considered a mental skill, but training for speed reading involves considerable effort in learning proper eye movement. Conversely, learning to ride a bicycle involves much more than turning the pedals and moving the handlebars. The student is successful only when the handlebars are turned exactly the right amount in response to eye and balance sense perceptions.

Desire to learn has a strong influence on the rate of learning. For example, when learning to ride a bicycle, youngsters who visualize the result in terms of increased mobility and joining their friends learn quite rapidly. However, if the bike is introduced to a matron as a form of exercise and she sees it as a threat in the form of exertion and an exposure to ridicule because of her clumsiness on it, success is measured in terms of the minimum time required to demonstrate that she can't possibly learn to ride.

The best way to prepare students to perform a given task is to provide them with a clear, step-by-step pattern to follow. In flight training, this usually is accomplished verbally during the preflight briefing, followed by a demonstration in the air. Whatever the method used, students need a clear impression of what they are to do.

In many cases, however, it is extremely difficult to tell someone else how to do something. Neither talking nor demonstrating seems to be of any help. All the learner can do is to keep trying, again and again, until suddenly, almost as if a switch is moved, the basic skill is achieved; the coordination circuit between the muscles and the perceptive senses is completed. Another benefit of practice is that, as the learner gains in proficiency, verbal instructions and demonstrations mean more. Where discussion and demonstration were once meaningless, a short comment may be very meaningful as the student begins to master the skill.

To maintain interest and provide incentive to continue with the training, students need to know how well they are doing. In some skills, this is quite apparent but, in others, mistakes may not be as obvious. Students may be aware that something is wrong, but may not know how to correct it. In either case, a helpful instructor is alert to the students' needs. It is just as important for students to know when they are right as when they are wrong. They should be informed as soon as possible, so they do not continue to practice improper techniques. According to the law of primacy, it is much more difficult to unlearn a mistake and then relearn the correct procedure than it is to learn correctly in the first place. A good method of informing students of their progress is for the instructor to redemonstrate the procedure and emphasize the standard with which they can compare their performance.

LEARNING CURVE

Laboratory experiments on skill learning consistently prove that the first trials are slow and coordination is lacking. Mistakes are frequent, but provide clues for improvement on further trials. A typical graph of the progress of skill learning is shown in

figure 2-3. Errors decrease rapidly during the early trials, but then the curve levels off and may stay nearly level for several practice periods. This leveling off is called a learning plateau and may result from any of a number of causes. Students may have reached the limit of their capabilities, their interest may have waned, or they may need more efficient instruction before they can progress further. It must be emphasized that the plateau does not necessarily mean that learning has ceased or that further progress is impossible. This leveling off process is to be expected after the normal initial period of rapid improvement. Students may feel frustrated and discouraged by this apparent lack of progress unless they are told to anticipate it.

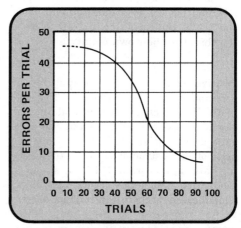

Fig. 2-3. Typical Learning
Curve for Motor Skills

A primary consideration in planning for student performance is the length of time devoted to practice. A beginning student quickly reaches a point where additional practice may reverse progress. To prevent discouragement, practice of the particular task should be terminated before tension and fatigue occur. Then, after the student has gained experience, longer practice periods will be profitable.

Another consideration is whether to divide the allotted time into segments or to plan on one long, continuous sequence. This depends largely on the student's stage of training. The process of learning to fly an airplane consists of related subgroups of skills which allow the instructor a certain amount of flexibility in planning a flight training session.

In the final analysis, the question is, "Can the student *use* the skill?" For the answer to be affirmative, two conditions must be present: the student must have learned the skill so well that it is easy, even habitual, to perform and there must be recognition of the situations where it is appropriate to use the skill. The second condition involves the transfer of learning, which is discussed later in this chapter.

PERMANENCY OF LEARNING

To be useful, the thing which was learned must be retained, so it can be recalled as needed. Consideration of why people forget may indicate ways to help them remember.

THEORIES OF FORGETTING

Disuse

One theory indicates that people forget those things which they do not use. Graduates of a school are dismayed by the small amount of factual information they retain several years after graduation. They remember only what they use frequently. A logical conclusion is that disuse has caused them to forget, but the answer is not that simple. Tests have shown that under hypnosis, people can recall facts which they could not recall in the normal conscious state. It seems, then, that the information is there but the ability to recall it has suffered from disuse.

Interference

Another theory indicates that material is forgotten because it is overshadowed by a similar or later experience. From experiments, two conclusions about interference may be drawn: closely similar material seems to interfere with memory more than dissimilar material; and, material not learned well suffers most from interference.

Repression

Freudian psychology advances the view that some forgetting is due to the submersion of ideas into the subconscious mind. Material that is unpleasant or which produces anxiety may unintentionally be treated this way.

RETENTION OF LEARNING

All of these theories indicate that when a person "forgets" something, it is not actually lost, but simply not available for recall. The instructor's job, then, is to exercise and stimulate the student's ability to recall what has been learned. The following suggestions may aid in this endeavor.

1. **Teach thoroughly and with meaning.** The more thoroughly something is learned, the less likely it is to be forgotten. This fact is supported by experimental studies. Meaningful learning, involving principles and concepts already a part of the student's own experience, becomes a part of those concepts, while rote learning is not easily retained because of its superficial nature.

2. **Help the student understand the reason for remembering.** A desire to remember increases the chances of remembering. A young man who normally cannot remember names is not likely to forget the name of an attractive young lady.

3. **Encourage the student to develop good study habits.** The student should be taught to study for meaning, to attempt to relate what is being learned to something already known. There should be encouragement and assistance in learning not only the "what" of the

subject, but also the "how" and "why." In this manner, if rote memory fails, it will still be possible to reason out the answer.

4. **Help the student develop an organized program of review.** Short periods of review, recitations, and quizzes are helpful, as well as periods of practice. Besides exercising the student's power of recall, occasional quizzes allow the instructor to evaluate the student's progress. Quizzes are especially valuable when they are designed to test the student's *application* of the knowledge gained. The ability to apply knowledge to a situation is the best proof of understanding.

TRANSFER OF LEARNING

During any learning experience, some things in a student's background will help and others will hinder. This is called transfer of learning, and it can be expressed as either a positive or negative factor. An example of positive transfer is the help that previous practice of slow flight and stalls gives to the student learning landings. Negative transfer is very apparent when teaching pilots of fixed-wing airplanes to fly a helicopter. Fixed-wing training conditions pilots to never let airspeed get too low except when over the runway, yet the helicopter instructor tells students to reduce airspeed to zero while still 5 or 10 feet in the air. In addition, airplanes need right rudder at high power settings and low airspeed; helicopters need left rudder. The result is that, in the earlier periods of training, it is much more difficult to teach the pilot of a fixed-wing airplane to fly a helicopter than it is to teach a person who has never flown before. As soon as positive transfer of learning can be employed, however, such as in ground reference maneuvers, the airplane pilot's rate of learning is much faster than the beginner's.

A negative transfer that is not so obvious may occur when some bit of academic knowledge a student is trying to assimilate strongly conflicts with something previously believed to be true. Due to the law of primacy, the instructor must convince this student to adopt a different belief. The instructor must probe deeply to uncover the reason for the student's reluctance to accept the information. This indicates a need to know as much as possible about the student's past experiences and current knowledge. The instructor who continually tries to tie new knowledge to old and new skills to those the student already possesses is providing maximum assistance to the student. Educational psychologists suggest the following steps to accentuate positive transfer.

1. Plan for transfer as a primary objective.
2. Make certain the student understands that what is learned can be applied in other situations. Encourage the student to search for other applications.
3. Insure thorough learning. The more thoroughly the student understands the material, the more apparent will be the relationship to new situations. Discourage rote learning; it does not foster transfer.
4. Provide learning experiences that challenge the student to exercise imagination and ingenuity in applying knowledge and skills.

FORMATION OF HABITS

Habits are formed by turning awareness into perceptions and combining perceptions into insights. Transfer is the application of performance habits formed in one task to the performance of more complicated tasks. Habits of significant importance in flight training are rudder and aileron coordination to prevent skidding and pitch attitude and power coordination to gain desired airplane performance.

Transfer of learning occurs when these small habits are employed to accomplish a subsequent task.

A flight syllabus usually is organized to take maximum advantage of positive transfer by introducing maneuvers in an order which utilizes the elements learned in one maneuver in the performance of subsequent maneuvers. The formation of correct habit patterns from the beginning of any learning process is essential to further learning. For example, scanning instruments during flight can be developed during an early stage to such an extent it is carried on without conscious attention. Because of this habit pattern, the pilot's attention is aroused immediately when something out of the ordinary is indicated. Pilots may remark, "I couldn't tell you why, but I just happened to notice that the oil pressure was low." They "happened" to notice it because it was their habit to scan the engine instruments. The habit was so ingrained that they were not even aware they were doing it as long as all indications were normal. The abnormal indications aroused attention, made them aware of the situation, and enabled them to take corrective action. If somewhere in their past they had not had an instructor who encouraged formation of this scanning habit, the first awareness of the low oil pressure condition could have been when the engine stopped.

The effects of good or bad habits are not always so obvious. If allowed to continue, poor coordination in the early stages of training becomes a habit a pilot must struggle to overcome later. Lack of attention to airspeed, altitude, or heading, if tolerated, can cause extra time later in the training program. As a student progresses, performance is either good or bad, depending on whether habit patterns are correct or incorrect. It is, therefore, the responsibility of the instructor to insist on correct procedures to provide proper habit patterns from the outset of training.

In the building-block technique of instruction, each simple task of instruction must be performed acceptably before the next task is introduced. Introduction of instruction in more advanced and complex operations before the initial instruction is mastered leads to the development and perpetuation of poor habit patterns which carry through to all future operations. Eventually, someone will have to take the time to discover and correct the faulty performance elements.

LEARNING OBSTACLES

An instructor must be aware of the many and varied obstacles to learning in order to overcome them. These obstacles may range from disinterest and distractions to complete mental blocks, and may originate from such diverse sources as family troubles or misconceptions carried over from previous instruction. Figure 2-4 depicts the major obstacles to effective flight instruction.

FEAR, ANXIETY, OR TIMIDITY

Fear, anxiety, or timidity place additional burdens on the flight instructor's ability to teach effectively. These are obstructions which limit the student's perceptive ability, and limit the formation of insights from the few perceptions which find their way into consciousness. The student must have confidence in the instructor and feel comfortable in the airplane. One of the most important tasks for the instructor is providing this atmosphere for learning. Instructors sometimes forget that a beginning student is entering for the first time into an environment that is taken for granted by the instructor. This affects some students more adversely than it does others.

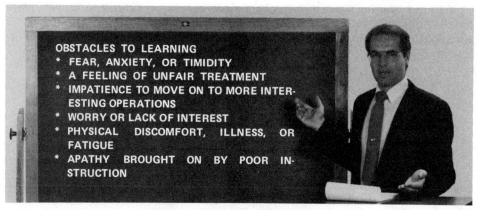

Fig. 2-4. Obstacles to Learning

Therefore, until the instructor is familiar with the student's reaction to the strange environment, it is advisable to avoid any maneuver which the student could interpret as threatening. As the student gains confidence in the situation, more drastic maneuvers may be introduced.

UNFAIR TREATMENT

Students who believe an instructor is not conscientiously considering and evaluating their efforts, do not learn well. If students feel that the instructor is not genuinely interested, motivation will suffer, no matter how much desire there is to learn to fly.

Motivation also suffers when students believe the instructor is making unreasonable demands for performance and progress. Assigning goals which are considered difficult, but attainable, provides challenges which students are eager to meet. However, if the attainment of the goals is impossible, there is a feeling of futility, and performance suffers.

IMPATIENCE

Impatience is a greater deterrent to learning than is generally recognized. It may take the form of the desire to make an early solo or a cross-country flight. Impatience often results from a failure to understand the need for the basic

instruction which necessarily precedes attainment of the ultimate goal. It can be curbed by clearly establishing the intermediate goals and identifying their significance in the overall training process.

A fast learner can experience impatience if instruction is keyed to the pace of the slow or average learner. It is just as important to advance a student to the subsequent step at the appropriate time as it is to continue drilling to insure readiness. Disinterest grows rapidly when unnecessary repetition and drill are conducted on operations which have been learned adequately.

WORRY, OR LACK OF INTEREST

Worry, or lack of interest for any reason, has a very detrimental effect on learning. A student who is emotionally upset derives little or no benefit from practice. Worry or distraction may be due to concern about progress in training or it may stem from circumstances completely unrelated to the instruction.

The student's experiences outside of training activities affect behavior and performance in training; the two cannot be separated. When students report for training, they bring with them all of their past experiences—their interests, enthusiasms, fears, and troubles. The instructor cannot be responsible for these outside diversions, but neither can they be ignored, because they vitally affect the conduct and results of teaching. Instruction must be keyed to the students' interests and enthusiasms, and an attempt must be made to divert the students' attention from those factors which would interfere wth learning.

Worries and upsets caused by the training situation can be remedied or prevented. They are usually evidence of inadequacies on the part of the instructor or the course itself. Discouragement and emotional upsets are rare when students feel they are part of their training and that nothing is being withheld or neglected in their training.

PHYSICAL DISCOMFORT

Physical discomfort, illness, and fatigue slow the rate of learning substantially in both the classroom and in-flight portions of the training program. When attention is diverted by discomforts such as temperature extremes, poor ventilation, inadequate lighting, or noise, perceptions are dulled and learning takes place at a slower rate.

Most illnesses adversely affect the acuteness of vision, hearing, and feeling. Since all of these are essential to the correct performance of a pilot's tasks, no effective flight instruction can be conducted when the student is incapacitated by illness.

Airsickness is a great deterrent to effective flight instruction. A student who is airsick or bothered with incipient airsickness is incapable of learning at a normal rate. There is no sure cure for airsickness, but resistance or immunity normally can be developed in a relatively short period of time if the flights are terminated at the first sign of airsickness. Then, as resistance is built

up, the length of the flights can be increased until normal training periods are practical. Repeated flights during which the student is allowed to become airsick, however, do not build up immunity. The instructor should keep the student interested and occupied during the entire flight to assist in the prevention of airsickness. A student is less likely to become sick while at the controls than when riding as a passenger. Tension increases susceptibility to airsickness so, until immunity is developed, maneuvers should be gentle.

Fatigue also results in a general downgrading in performance. Therefore, prevention of fatigue is the primary consideration governing the length and frequency of instruction periods. Individuals vary in the amount of training they can absorb without becoming fatigued. This capability is not based on either physical or mental stamina and, therefore, cannot be judged in advance. If the student is not alert and receptive to instruction or if the level of performance is not consistent with experience, it is reasonable to assume that fatigue has set in, and the instruction period should be altered or discontinued. When fatigue occurs as the result of application of a learning task, respite should be offered in the form of a break in instruction or a change of pace. Fatigue can be delayed by introducing a number of different maneuvers involving different elements.

APATHY

Apathy develops rapidly when the student realizes the instructor is inadequately prepared or when the instruction is deficient, contradictory, or appears insincere. To hold the student's interest and to maintain a high level of motivation, well planned, appropriate and accurate instruction must be provided. Even an inexperienced student realizes immediately when the instructor has failed to prepare for the lesson. Spotty coverage, misplaced emphasis, unnecessary repetition, and disorganized presentation can cause the student to lose confidence in the instructor. Poor planning also is evident when instructions are overly explicit and so elementary that they insult a student's intelligence or they are so general the student cannot make sense out of them. To make the experience relevant in the student's eyes, the presentation must be geared to the student's total experience.

Poor presentations of instruction may result not only from poor preparation but, also, from distracting mannerisms, personal untidiness, or the appearance of irritation with the student. Flight instructors normally work in rather close quarters and, therefore, should be attentive to those problems emphasized in the deodorant and mouthwash commercials.

FLIGHT TRAINING AND THE INSTRUCTOR

Learning to fly should be an enjoyable experience. By making each lesson a pleasant situation for the student, the flight instructor can maintain a high level of motivation. This should not be construed to mean that things must always be made easy for the student or that the instructor has to sacrifice standards to please the student. The student experiences pleasure from successful accomplishment of a learning task, just as surely as from a joy ride.

The idea that people must be led to learning by making it easy has no basis in fact. There is more self-enhancement to be gained from the successful outcome of a difficult asignment than from just sitting back and taking it easy. People want to feel capable; they are proud when they achieve difficult tasks. A good instructor can motivate

the student to tears and sweat, and the student will be grateful for it when the particular goal is achieved.

Learning to fly should be interesting. Instructors can sustain a student's interest only if they build up their own interest in the goal of each lesson. Knowledge of the objective of each period of instruction gives meaning and interest to the instructor's service and to the student's efforts. Not knowing the objectives involved leads to confusion, disinterest, and uneasiness on the part of the student, and it doesn't help the instructor's task, either.

Flight training should provide an opportunity for exploration and experimentation. Students should be allowed time to try things for themselves and to evaluate the various elements of each maneuver or operation presented. They must discover their own capabilities and build self-confidence. They must be given enough latitude to make mistakes, so they can perceive the outcome.

Early training should consist of habit-building periods during which students devote their attention, memory, and judgment to the development of correct habit patterns. Any goal other than a desire to learn the right way makes students impatient with the instruction and practice they need and should be trying to obtain. Instructors must keep this goal before them by examples and logical presentations of learning tasks. When flight instructors understand the specific objectives they seek and can employ the principles which govern the learning process, they can teach these objectives efficiently.

BEHAVIOR

The behavioral sciences are rich in knowledge of the management of people and the use of human resources. This is a valuable source of knowledge for instructors, who may do a disservice to their students if they try to rely exclusively on their personal experiences and observations to create the proper training atmosphere. Establishment of good relations with students means far more than just getting along. One of the instructor's major tasks is organizing their students' activities to promote a climate conducive to learning. This can be done successfully only by applying the ability to predict and control the students' behavior.

CONTROLLING BEHAVIOR

The relationship between instructors and students has a profound impact on how much students learn. Students perceive their instructors as the symbols of authority during the time they share the classroom and the airplane. Students expect instructors to exercise certain controls, and they recognize and submit to authority as a valid means of control. The instructor's challenge is to know what controls are best for the particular circumstances.

Students work toward goals that are peculiar to their own values, beliefs, and interests. Whatever the goals, instructors are obligated to help their students achieve them by directing and controlling behavior and by guiding them toward their goals. Without the instructor's active intervention, students would be passive and perhaps resistant to learning. The amount and type of control instructors exercise should be based on more than trial and error or on their feelings at the moment. To mold solid, productive relationships with students, instructors must understand the desires and needs that human beings are constantly striving to satisfy. Some of those needs and drives are discussed in the following paragraphs.

Needs

Superior instructors never lose their awareness of the fact that students are human beings. The needs of students generally can be organized by their level of importance, as illustrated in figure 2-5.

At the bottom of the list is the broadest, most basic category, the physical needs. Each person is first concerned with a need for food, rest, exercise, and protection from the elements. Until these needs are satisfied, a person cannot concentrate fully on any other task at hand. Once a need is satisfied, it no longer provides motivation, because a want that is satisfied is no longer a want. Thus, the person strives to satisfy the needs of the next higher level.

The safety needs are protection against danger, threat, and deprivation, labeled by some as the security needs. No matter what they are called, they are very real, and a student's behavior is influenced by them.

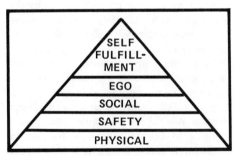

Fig. 2-5. Student Needs

When individuals are physically comfortable and do not feel threatened, they seek to satisfy their social needs. These are to belong, to associate, and to give and receive friendship and love. Since students are usually out of their normal surroundings during flight training, their need for association and belonging will be more pronounced. Flight school personnel should make every effort to help new students feel welcome and at ease.

The egoistic needs usually have a strong influence on the instructor-student relationship. Those needs are of two kinds: those that relate to one's self-esteem, such as self-confidence, independence, achievement, competence, and knowledge; and the needs that relate to one's reputation, such as status, recognition, appreciation, and respect of associates.

At the apex of the pyramid is self-fulfillment—realizing one's own potentials for continued development and creativity in the broadest sense of the term. These needs offer the greatest challenge to instructors, and assisting in their fulfillment brings the greatest rewards.

DEFENSE MECHANISMS

When someone or something prevents people from achieving their goals, they become frustrated, and cannot give their full attention to learning. People react to frustration in many ways. By observing the reactions of students, instructors can recognize their behavior patterns and identify those whose problems interfere with their work.

Certain behavior patterns are called *defense mechanisms,* because they are subconscious defenses against the realities of unpleasant situations. People use these defenses to soften feelings of failure, alleviate feelings of guilt, and protect feelings of personal worth and adequacy.

Although defense mechanisms can serve a useful purpose, they also can be hindrances. Because they involve some self-deception and distortion of reality, defense mechanisms do not solve problems, just as a pain reliever alleviates symptoms, rather than curing the causes. Common defense mechanisms are rationalization, flight, aggression, and resignation.

Rationalization

If students cannot accept the real reason for their behavior, they may rationalize. This allows them to substitute excuses for reasons; moreover, they can make those excuses plausible and acceptable to themselves. Rationalization is a subconscious method for justifying an action which is otherwise unacceptable, and the individuals sincerely believe their excuses are real and fully justifiable. A moderate amount of rationalization is normal, but there is potential danger when rationalization is used excessively or as a means of excusing inadequate performance.

Flight

An individual often escapes from a frustrating situation by taking flight, either physically or mentally. To take flight physically, the person may develop symptoms or ailments which provide a satisfactory excuse for getting away from the scene of the frustration. In such cases, the symptoms are subconsicous reactions, not a conscious effort. Although the ailment may be mental rather than physical, the symptoms are very real and, as such, can be painful.

Mental flight, or daydreaming, is more frequent than physical flight. Mental flight provides a simple and satisfying escape from an uncomfortable situation. Those who get enough satisfaction from daydreaming may abandon their goals altogether. Carried to extremes, fantasy and reality become difficult, even impossible, to distinguish. The dreams of what students want to be cancel out the knowledge of what they are. A certain amount of daydreaming is normal, but students who daydream in class miss the opportunity to learn.

Aggression

Anger is a normal universal emotion. Angry persons may shout, swear, slam a door, or give vent to the heat of their emotions in a number of different ways. In a classroom, however, such extreme behavior is infrequent. Because of social restrictions, aggressiveness in the classroom usually takes a more subtle form. Students may ask irrelevant questions, refuse to participate in class activities, or disrupt activities within their group. If they cannot deal directly with the cause of their frustration, they may vent their feelings on a neutral object or person not even related to their problem; or they may turn inward to the extent they are blinded to what is going on around them. When this happens, of course, they cannot learn.

Resignation

Students may become so frustrated that they lose interest and give up; they are no longer convinced that it is profitable or even possible to achieve goals. The most obvious cause for resignation is the inability to master the material presented. Sometimes resignation is brought on when, after completing an early phase of a course without grasping the fundamentals, students become bewildered in the advanced phase. From that point on, their learning is negligible, although they may continue to go through the motions of participating.

HUMAN RELATIONS AND THE INSTRUCTOR

It is equally as important for instructors to develop and apply skill in human relations as it is for them to apply their skill in all the aspects of flying. There are a few guidelines which can help instructors minimize the frustrations of their students and achieve good human relations.

KEEP STUDENTS MOTIVATED

Students gain more when they want to learn than when they are forced to learn. They will want to learn when they realize how a particular lesson or course can help them reach an important goal. Use of a clearly defined objective, plus the conviction that the objective is a step toward this goal, is the best way to accomplish this purpose.

KEEP STUDENTS INFORMED

Students can feel insecure when they do not know what is expected of them or what is going to happen to them. Instructors can minimize such feelings by telling students what is expected from them and what they can expect. This may be accomplished in various ways—giving them an overview of the course, keeping them posted on their progress, and giving them adequate notice of examinations, assignments, and administrative requirements.

CONSIDER STUDENTS AS INDIVIDUALS

When instructors consider a group without recognizing the individuals who make up the group, they are directing their efforts toward an average personality which really fits no one. Although a group may have its own personality, stemming from the characteristics and interactions among its members, each individual within the group has a personality which is unique and which must be considered.

GIVE CREDIT WHEN DUE

When a person does something very well, that person wants to have those abilities and efforts noticed; otherwise, they become frustrated. Praise or credit from instructors usually is ample reward, and provides an incentive to do even better. Praise given too freely, like any other commodity in overabundant supply, is cheap; but when it is deserved, it pays dividends in student attitude and achievement.

CONSTRUCTIVE CRITICISM

The key word concerning criticism is "constructive." Although it is just as important to identify mistakes and failures as it is to give praise, telling students they have made a mistake without telling them why it occurred or how it may be overcome does not help. Frustration is the inevitable result if students have done their best, but are told that the work was unsatisfactory, with no other explanation. They cannot correct their errors if they do not know what they are, and these uncorrected errors may develop into habits. On the other hand, if they know exactly what they did wrong and how to go about correcting it, they can progress and enjoy a feeling of accomplishment.

BE CONSISTENT

Students want to please their instructors, and will have a keen interest in knowing what is required to please them. If the same performance appears to be acceptable one day and unacceptable the next, students become confused. Instructors must, therefore, be consistent in philosophy and actions.

ADMIT ERRORS

No one, not even students, expect instructors to be perfect, and they should not try to project that image. Instructors can win the respect of their students by honestly acknowledging their mistakes, rather than trying to cover up or bluff. If in doubt, they should admit it to their students, then give them the correct information at the earliest opportunity.

All professional instructors are concerned with many areas of learning in addition to the area in which they teach. A particular area which requires extensive knowledge is that of the behavioral sciences. If students are to learn effectively, instructors must insure that they are prepared to receive knowledge by stimulating their desire to learn. Students' motivation must be maintained at the highest possible level so knowledge grows into perceptions, perceptions combine into insights, and, from the formulation of insights, students achieve the ability to apply what they have learned. The application of good human relations skills is one of the instructor's best methods for keeping students in a frame of mind conducive to learning, and for

eliminating doubt, frustration, and discouragement.

EFFECTIVE COMMUNICATIONS

The ability to communicate effectively is essential for a successful instructor. Communicating effectively, like most human endeavors, is complex, variable, and at times, uncertain. Communicating as an instructor is even more difficult because of the variations and complexities of the teaching-learning process. In a technical field such as aviation, an additional difficulty is encountered as a result of the different vocabularies used by the instructor and the student. The process of communication can be examined by analyzing its cardinal elements and the significant relationships between them in order to gain a deeper, more precise understanding of the process.

BASIC ELEMENTS

Communication takes place when one person transmits ideas or feelings to another person or group of people. Its effectiveness is measured by the similarity between the idea transmitted and the idea received.

The process of communication is composed of three elements: the *source* (a sender, speaker, writer, instructor, or transmitter), the *symbols* used in composing and transmitting the message (words, signs, music), and the *receiver* (a listener, reader, or student). These elements are dynamically interrelated; that which affects one influences the others. This relationship is not only dynamic, but is also reciprocal—it is a two-way process.

SOURCE

The communicator's effectiveness is related to at least three basic factors. First, an ability to select and use language influences intelligibility to listeners and readers. Second, communicators consciously or unconsciously reveal their attitudes toward themselves as a communicator, toward the ideas they are trying to communicate, and toward their receivers. These attitudes must be positive if they are to communicate effectively. They must be confident. They must indicate they believe their message is important and make it clear they believe their listeners have a need to know their ideas. Third, successful communicators speak from a broad background of accurate, up-to-date, stimulating material. They must exercise great care to make certain their ideas and feelings are meaningful to their listeners. Far too often, a speaker or a writer may depend on a highly technical or professional background, with its associated vocabulary that is meaningful only to others with a similar background. Reliance on technical language often impedes effective communication.

SYMBOLS

At its basic level, communication is achieved through the use of simple oral and visual codes. The words in the vocabulary constitute a basic code. Common gestures and facial expressions form another, but words and gestures alone do not communicate ideas. They must be combined into units which mean something to the student—sentences, paragraphs, lectures, or chapters. When combined into these units, each portion of the whole becomes important to the total effort.

The parts of the total idea must be analyzed to determine which are most suited to starting or ending the communication, and which are best for the purpose of explaining, clarifying, or emphasizing. All of these functions are required for effective transmission of ideas.

The process finally culminates in the determination of the medium best suited for their transmission. Most frequently, communicators select the channels of hearing and seeing. For motor skills, the sense of feeling is added as the student practices the skill. The most successful communicators use a variety of channels for transmitting their carefully selected ideas.

RECEIVER

Effective communicators always remember an overall basic guideline—communication succeeds only in relation to the reaction of the students. When students react with understanding and change their behavior according to the intent of the sender, then, and only then, has effective communication taken place.

To understand the process of communication, at least three characteristics of students must be understood—their abilities, attitudes, and experiences. First, they exercise their ability to question and comprehend the ideas that have been transmitted. Instructors can capitalize on this ability by providing an atmosphere which encourages questioning. Second, the students' attitude may be one of resistance, willingness, or passive neutrality. To gain and hold the students' attention, attitudes must be molded into those which promote reception. The more the communicative approach is varied, the greater the success in reception. Third, the students' background, experience, and education frame the target at which instructors must aim. The students' knowledge must be assessed as the fundamental guide for the selection and transmission of ideas. The major barriers to communication are usually found in this particular area.

- LACK OF A COMMON CORE OF EXPERIENCE
- CONFUSION BETWEEN THE SYMBOL AND THE OBJECT SYMBOLIZED
- EXCESSIVE USE OF ABSTRACT TERMS

Fig. 2-6. Barriers to Effective Communication

BARRIERS TO EFFECTIVE COMMUNICATIONS

The nature of language and the way it is used often lead to misunderstandings. A classic example is the pilot who, wishing to make a go-around, calls for "takeoff power," and the copilot dutifully *takes off* power. As illustrated in figure 2-6, misunderstandings stem primarily from three barriers to effective communications.

LACK OF COMMON EXPERIENCE

Probably the greatest single barrier to effective communication is a different background of experience between the instructor and student. Many people seem to believe that words transport meanings from speaker to listener in the same way that a truck carries bricks from one location to another. Words, however, rarely carry precisely the same meaning from the mind of the instructor to that of the student. In fact, words, in themselves, do not transfer meanings at all. Whether spoken or written, they are merely stimuli used to arouse a response in the receiver. The nature of the response is determined by the receiver's past experiences with the words and the things to which they refer. Since it is the students' experiences that form vocabulary, it is essential that instructors insure they are "speaking the same language" as their students. If the instructor's terminology is the only thing that can convey the idea, some time must be spent to insure the students understand it.

CONFUSION BETWEEN THE SYMBOL AND THE SYMBOLIZED OBJECT

The English language abounds in words which mean different things to different people. To a farmer, the word "tractor" means the machine that pulls the implements to cultivate the soil; to a trucker, it is the vehicle used to pull a semitrailer; in aviation, a tractor propeller is the opposite of a pusher propeller. Each technical field has its own vocabulary—words which mean something entirely different to a person outside that field, or perhaps mean nothing at all.

Although it is obvious that words and the connotations they carry can be different, people sometimes fail to make the distinction. A man may lose his job because someone calls him a thief. If people act as though being called a thief and *being* a thief are the same, they are confusing a word with the thing it represents. On the assumption that many people will "buy a label," or accept a word for being a thing, manufacturers invest much money in naming their products. Effective speakers and writers carefully differentiate between symbols and the things they represent, keeping both in perspective.

OVERUSE OF ABSTRACTIONS

Concrete words or terms refer to objects that human beings can experience directly. They specify an idea which can be perceived or a thing which can be visualized. Abstract words, on the other hand, stand for ideas that cannot be directly experienced, things that do not call forth mental images in the minds of the receivers. The word "aircraft" is an abstract word. It does not call to mind a specific aircraft in the imaginations of various students. One may visualize an airplane, another a helicopter, and still another may imagine a blimp. Although the word "airplane" is more specific, the imagination may call up anything from a Piper Cub to a Boeing 727.

Abstractions are necessary and useful. The Cherokee isn't exclusive in its reaction to the four forces acting on an airplane in flight, so generalizations can be helpful during ground training. Aerodynamics, another abstract term, is applicable to all aircraft. But abstractions can lead to understanding. The danger is that they will not invoke the specific items of experience in the listener's mind that the communicator intends. Such terms as "proper measures" and "corrective action," frequently heard in flight schools, may mean nothing to a student or, even worse, may invoke the thought of the wrong procedure. When such terms are used, they should be linked with specific expriences through examples and illustrations. Whenever possible, the level of abstraction should be reduced by using concrete and specific terms as much as possible to narrow and gain better control of the image produced in the mind of the listener.

An awareness of the three basic elements of the communicative process—the source, the symbols, and the receiver—indicates the beginning of the understanding required of the successful communicator. Recognizing the various characteristics of each element and using this recognition as a basis for increased understanding can help an instructor overcome inherent barriers in transmitting ideas and feelings.

SECTION B—THE TEACHING PROCESS

There are four basic steps in the teaching process that are essential to effective instruction. These steps, shown in figure 2-7, may be defined in different terms, listed in different order, or broken down in greater detail, but they are always recognized in any serious consideration of the teaching process.

PREPARATION

The instructor's lesson plan may be prepared mentally, in the case of an experienced instructor planning a simple period of instruction, or it may be worked out with care and prepared in written form. The lesson plan is simply a statement of the lesson objectives, the specific goals to be attained, the procedures and facilities to be used, and the means of evaluating the desired results.

The critical point in the lesson plan is determination of the objective. In order to accomplish the task in the most efficient manner, the instructor must remember that the objective must be stated in terms of observable behavior, the conditions under which the behavior will occur, and the minimum acceptable performance.

The minimum performance limits do not need to reflect the final goal. A student performing a maneuver for the first time should not be expected to perform as well as one experienced in the maneuver. The goals for the particular period should be attainable, so the student can experience success, rather than frustration at not meeting goals which were unattainable at the present experience level.

PRESENTATION

As stated earlier, the learning process begins with awareness, proceeds to perceptions which combine into insights, then continues until the insights are formed into habits. In learning a foreign language, for example, the student first becomes

- PREPARATION
- PRESENTATION
- APPLICATION
- REVIEW AND EVALUATION

Fig. 2-7. Basic Steps in the Teaching Process

aware of a word, then perceives its meaning and begins to understand its use. Final proof that the student has learned the word is an ability to recall it automatically and apply it in the proper place to express an idea. In motor skills, such as flying, the student becomes aware of the results of certain control applications, gains insight into why the airplane reacts as it does, then practices to form habits. The habits are transferred and combined with other habits to perform subsequent maneuvers.

The presentation step, then, is the one in which the student is introduced to the subject of the instruction period. The instructor is the most active during this step. In the case of a flying lesson, it may be divided into a preflight briefing and an in-flight demonstration. At this stage, the instructor employs skills in communication to insure the student thoroughly understands what is to be accomplished during the instruction period and what will demonstrate mastery of the subject. The student also should understand the relative importance of the lesson's subject in the overall scheme of things.

APPLICATION

During the application step, through trial and practice, the student is guided into development of insights and formation of habits. The student is now the active person, with the instructor observing and correcting or demonstrating further, as required. In classroom instruction, this step may consist of recitation or problem solving. In flight, the student is practicing the maneuver or operation which has been explained or demonstrated.

Although they are technically separate segments of the lesson, portions of the instructor's demonstration and explanatory activity are usually alternated with portions of the student's trial and practice activity. It is rare that the instructor completes the explanations and demonstrations, then allows the student to accomplish the trial and practice activities without interruptions for corrections and further demonstrations.

REVIEW AND EVALUATION

Before concluding the instruction period, it is vital that the instructor summarize what has been covered during the lesson, and require the student to demonstrate the extent to which the lesson objectives have been met. The instructor's evaluation may be informal and noted only for its use in planning the next lesson, or it may be recorded to certify the student's progress.

In either case, the student should know what progress has been made. Failure of the instructor to insure the student is cognizant of the progress or lack of progress may impose a barrier between them. The instructor must remember it is difficult for a student to obtain a clear picture of the progress made, since there is little opportunity to compare it with the progress of others who have a similar experience level. The student can make direct comparisons only with the performance of the flight instructor, which tells very little about the performance in comparison with other students with similar backgrounds. Only the instructor can provide a realistic evaluation of a student's performance and progress, information that is vital if motivation is to remain at a high level.

In addition to knowledge and skills learned during the instruction period just completed, each lesson should include a review and evaluation of things learned previously. If a deficiency or fault is revealed that pertains to factors concerning the present lesson, it must be corrected before the new lesson can begin. If deficiencies not asociated with the present lesson are noticed, they should be noted carefully and pointed out to the student. Corrective measures, which are practical within the limitations of the situation, should be taken immediately and more thorough remedial actions must be included in future lesson plans.

The evaluation of the student's performance and accomplishments during the lesson should be based on the objectives established in the lesson plan. Evaluations for the purpose of certifying the completion of stages of training or for executing student pilot certificate endorsements should be based on the standards of the appropriate training syllabus.

TEACHING METHODS

The preceding discussion has been basically theoretical, concerning areas of knowledge pertinent to all activities involved in the training process. The development of habits based on this knowledge and skill in the application of the principles discussed will enable instructors not only to approach their certification examinations with confidence, but also will enable them to conduct their teaching activities more efficiently and successfully. The discussion which follows departs from the theoretical with some specific recommendations concerning the actual conduct of the teaching process, to enable instructors to select procedures which have been tested and found effective.

ORGANIZING MATERIAL

When the lesson objectives have been decided upon and stated in terms of specific behavior, and the necessary course material has been gathered, the lesson should be organized to develop and support the learning outcomes. One of the most effective ways to organize a lesson is to divide it into parts. A frequently recommended division breaks the lesson into an introduction, a development, and a conclusion.

INTRODUCTION

The introduction should serve several purposes. It should establish common ground between the instructor and students, capture and hold their attention, indicate what is to be covered and relate it to the entire course, point out specific benefits the student can expect from the learning, and establish a receptive attitude toward the subject. In brief, the introduction sets the stage for learning.

Attention

The instructor might begin by telling a story that relates to the subject and establishes a background for developing the learning goals. Attention might be gained by making an unexpected statement or by asking a question that helps students relate the lesson topic to the welfare of the group. Whatever method is used, the main concern should be to gain the attention of the students and to focus it on the subject.

Motivation

The introduction should offer specific reasons for needing whatever the students are about to learn. This motivation should appeal to each student personally. The appeal may relate the learning to career advancement, financial gain, service to community groups, use at home, or some other attraction. In every instance, a specific application should be cited.

Overview

Every lesson introduction should contain an overview that tells the students what is to be covered during the period. A clear, concise presentation of the objective and how well it is to be performed gives the students a clear view of the route to be followed. The introduction should be free of stories, jokes, or incidents that do not help the students focus attention on the objective.

DEVELOPMENT

Development is the main part of the lesson. Here, the instructor develops the subject matter in a manner that helps the students achieve the desired outcomes. The instructor must organize the material logically to show the relationships of the main points. These relationships usually are shown by developing the main points in one of the following ways—from the past to present, from the simple to the complex, from the known to the unknown, from the most frequently used to the least frequently used.

Past to Present

In the past to present pattern of development, the subject matter is arranged chronologically. Such time relationships are most suitable when history is an important consideration, as in tracing the development of aircraft design.

Simple to Complex

Simple to complex development helps the instructor lead the student from simple facts or ideas to an understanding of involved concepts or procedures. In studying jet propulsion, the student might begin by considering the action involved in releasing air from a balloon and finish by taking part in a discussion of a complex gas turbine.

Known to Unknown

By using something the student already knows as the point of departure, the instructor can lead into new ideas and concepts. In developing the theory of flight, for example, many instructors begin by using the simple illustration of putting a hand out of the window of a moving car or drawing a parallel to the action of a kite.

Most Frequently Used to Least Frequently Used

In some subjects, certain information or concepts are common to all who use the material. This fourth organizational pattern starts with common usages before progressing to the rarer ones. In a course on navigational proficiency, students may start by learning dead reckoning procedures, which become the basis for their study of other forms of navigation, such as VOR or RNAV.

Under each main point in a lesson plan, the subordinate points should lead naturally from one to the other. Meaningful transitions from one main point to the next keep the students oriented and aware of where they have been and where they are going.

CONCLUSION

The conclusion should include review, remotivation, evaluation, and closure. The review retraces the important elements of the lesson and relates them to the objective in order to reinforce the students' learning and help them retain what they have learned. The purpose of remotivation is to instill in the students a desire to make use of what they have just learned. The evaluation, whether formal or informal, informs the students of their progress. The closure may be a quotation, a concise statement, or any other device which will serve as a conclusion. The closing statement may be designed to incorporate the remotivation step.

LECTURE METHOD

When organization of the material is complete, the method of presentation is the next decision to be made. One of the methods frequently used is the lecture method. Although the trend in education is away from the lecture and toward more involved activity for the student, certain times and situations still point to the lecture as the most suitable instructional approach. It is used primarily to introduce new material, but it also is a valuable method for summarizing ideas, showing relationships between theory and practice, and reemphasizing main points.

TYPES OF LECTURES

Oral presentations take many forms. Of the various approaches, instructors seem to favor the four illustrated in figure 2-8.

ILLUSTRATED TALK

In the illustrated talk, the speaker conveys most of the main ideas through the use of visual aids. Speech is used merely to describe or clarify the ideas the visual aids introduce. Audiovisual programs are a more formal version of the illustrated talk method. In the audiovisual presentation, the instructor needs only to amplify the points not fully understood by the students by answering their questions.

BRIEFING

In a briefing, the speaker presents an array of significant facts as concisely as possible. Details are avoided unless they are critical and necessary. The briefer provides supplementary material only if asked to do so. The preflight briefing exemplifies this type of presentation. The instructor explains what will be done on the flight, the maneuvers to be reviewed, new maneuvers to be introduced, and what is to be expected of the student. In a briefing, the instructor does not attempt to develop basic knowledge or understanding, but only to inform.

FORMAL SPEECH

The formal speech generally is designed to inform, persuade, or entertain. Each type of formal speech has its own special characteristics.

The *speech to inform* is a narration concerning a particular topic. This type of speech usually is not related to any sustained effort to teach something. A travelogue is an excellent example of a speech to inform.

The *speech to persuade* is designed to move an audience to take action pertinent to some topic, product, way of life, or other matter. Like the informative speech, the speech to persuade is generally an isolated event that is not related to any sequence of oral presentations. Recruiting speeches

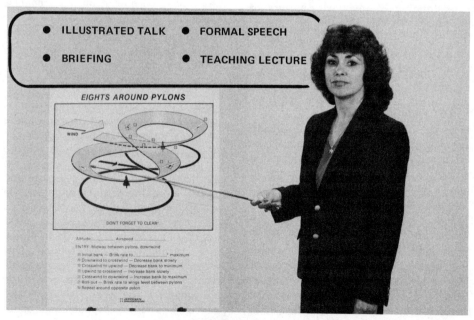

Fig. 2-8. Types of Lectures

to a graduating class, political speeches, and sermons are examples of persuasive speeches.

The purpose of the *speech to entertain* is to give enjoyment to the members of the audience. The speaker usually relies on humor and vivid language as a primary means of entertaining the listeners. The after dinner speech is frequently a presentation of this type. While speeches to inform, persuade, or entertain serve useful purposes, they are not designed to further the education of those who hear them and, therefore, are of little or no use to an instructor.

TEACHING LECTURE

In the teaching lecture, instructors are not only responsible for planning, presenting, and supporting the lesson, but also for the direction and depth of ideas presented. They must determine if their ideas should progress from the known to the unknown or from the simple to the complex, and to what extent the ideas will be covered.

In other methods of teaching, instructors receive direct reaction from the students in the form of verbal or motor activity. In the teaching lecture, however, the feedback is not as direct and is much harder to interpret. In this type of lecture, instructors must develop a keen perception for subtle responses from the class, such as facial expressions, manner of taking notes, and apparent interest or disinterest in the lesson. They must be able to interpret the meaning of these reactions and adjust their lessons accordingly.

PREPARING THE TEACHING LECTURE

Competent instructors know that careful preparation is one key to successful teaching performance. This preparation should follow the four steps outlined under the heading of "The

Teaching Process." In all stages of preparing for the teaching lecture, instructors should support any point to be covered with examples, comparisions, statistics, or testimony. They must consider the nature of their audience and work on the assumption that they may neither believe nor understand the points to be covered without the use of supporting information.

Suitable Language

Simple, rather than complex words should be used in the classroom lecture whenever possible. Good newspapers offer examples of the effective use of simple words. Picturesque slang and free-and-easy colloquialisms which suit the subject can add variety and vividness to a classroom lecture; however, the instructor should not use substandard English. Vulgarisms and errors in grammar detract from an instructor's dignity and reflect upon the intelligence of the students.

If the subject matter includes technical terms, each one should be defined clearly so that the students are not in doubt about their meanings. Whenever possible, instructors should use specific, instead of general words. For example, the specific words "a leak in the gas line" provide more information than the general term "mechanical defect."

Another good lecture technique is to use sentences of varying lengths. The consistent use of short sentences results in a choppy style. On the other hand, unless long sentences are carefully constructed, they are difficult to follow. When used poorly, long sentences can become too tangled for comprehension. To insure clarity and variety, instructors should combine use of sentences of short and medium lengths.

Types of Delivery

A speech can be delivered in one of four possible ways: reading from a manuscript, reciting memorized material, speaking extemporaneously from an outline, or speaking impromptu, without preparation. Although many speakers appear to favor the latter, the results are usually proportional to the preparation. The teaching lecture is probably best when delivered in an extemporaneous manner. With this method, instructors speak from a mental or written outline, but do not read or memorize what they are going to say. This gives the lecture a more personalized tone than one that is being read or spoken from memory. Since they talk directly to the students, instructors can readily observe their reactions and adjust to their responses. For example, if they realize from their expressions that a large number of students fail to grasp an idea, they can elaborate on the idea until their reactions indicate that they understand. The extemporaneous presentation reflects the instructor's personal enthusiasm and is more flexible than other methods, so it is more likely to hold the attention of the students.

Formal Versus Informal Lectures

The lecture may be conducted in either a formal or informal manner. The informal lecture includes active student participation. The primary goal in any teaching method is the achievement of objectives. Learning is best achieved if students participate actively in a friendly, relaxed atmosphere. Therefore, the use of the informal lecture is encouraged. At the same time, it must be realized that a formal lecture is preferable on some occasions, such as for introduction of new subject matter.

Active student participation can be achieved through the instructor's use of questions. Questions may be used to determine the experience and background of the students (so the lecture can be tailored to their needs), to add variety and stimulate interest, or to check student understanding.

The lecture is the most economical of all teaching methods in terms of the time required to present a given amount of material. To insure that all students have the necessary background to learn a subject, the instructor can present the basic information in a lecture. In this way, students with varied backgrounds can be offered a common understanding of principles and facts.

Although the lecture method can help the instructor meet the special challenges just discussed, it does have drawbacks. Too often the lecture does not provide for student participation and, as a consequence, many students will sit back and let the instructor do all the work. Learning is an active process, and the lecture method tends to foster passiveness. Many instructors find it difficult to hold the attention of all students in a lecture lasting throughout the class period. Because of this factor, an instructor needs considerable skill in speaking to achieve desired learning objectives through the lecture method.

THE GUIDED DISCUSSION METHOD

Because a guided discussion is particularly adapted to student participation, instructors should use it whenever possible, especially when dealing with small groups. In a group discussion, the student can fulfill personal needs by working with other students. The high morale and group consciousness resulting from cooperative effort stimulate the students to display initiative and to think creatively.

KINDS OF QUESTIONS

In the guided discussion, learning is produced through the skillful use of questions which can be categorized by functions and characteristics. Understanding these distinctions helps the instructor become a more skilled user of questions.

The instructor often uses a question to open up an area for discussion. This is the leadoff question, and its function is indicated by its name. After the discussion develops, the instructor may ask a followup question to guide the discussion. The reasons for using a followup question may vary. For example, it may be desirable for a student to consider an idea more deeply or to explain something more thoroughly, or there may be a need to bring the discussion back to a point from which it has strayed.

In terms of their characteristics, questions can be identified as overhead, rhetorical, direct, reverse, and relay. The overhead question is directed to the entire group to stimulate thought and response from each group member. The instructor uses an overhead question to pose the leadoff question. The rhetorical question is similar in nature, because it also spurs group thought. However, since the instructor answers the rhetorical question itself, it is more commonly used in lecturing than in guided discussion.

When the instructor wants to phrase a question for followup purposes, the overhead type may be chosen. However, if a response from a specific individual is desired, a direct question can be posed to that student. When a student asks a question, the instructor may respond with a reverse question. In this case, the instructor redirects the question so the student can provide an answer, rather than giving a direct answer to the student's query. A relay question is similar to a reverse question, except it is directed to the group instead of the individual.

PLANNING A GUIDED DISCUSSION

Planning for a guided discussion is basically the same as planning for a lecture. The following suggestions will be helpful in planning a discussion lesson.

1. Select a topic the students can discuss profitably. If necessary, make assignments that will give the students an adequate background for discussing the lesson topic.

2. Establish a specific lesson objective. Through discussion, the students develop an understanding of the subject by sharing knowledge, experiences, and backgrounds.

3. Plan at least one leadoff question for each desired learning outcome. In preparing questions, remember that the purpose is to bring about discussion, not merely to get answers. Avoid questions that require only short categorical answers, such as "yes," "no," "one," etc. Leadoff questions usually begin with "how" or "why."

STUDENT PREPARATION

The instructor should help the students prepare themselves for the discussion before coming to class. Each student should be encouraged to accept responsibility for contributing to and profiting from the discussion. The lesson objective should be emphasized throughout the time the instructor is preparing the students for their discussion. When there is no opportunity to assign preliminary work, it is practical and advisable for the instructor to give the students a brief general survey of the topic in the introducton. Under no circumstances should students without some background knowledge be asked to discuss a given subject.

GUIDING A DISCUSSION

Introduction

A guided discussion lesson is introduced in the same manner as the lecture. The introduction should include attention and motivation steps and an overview of key points. To encourage enthusiasm and stimulate discussion, the instructor should create a permissive atmosphere. Students should be given the opportunity to discuss the various aspects of the subject and should be made to feel that they have a personal responsibility to contribute.

Discussion

The instructor may open the discussion by asking a prepared leadoff question. After asking a question, the instructor should be patient while the students have a chance to react. Although the instructor has the answer in mind before asking the question, the students must think about the question before they can answer. The more difficult the question, the more time the students will need to produce an answer. In response to puzzled expressions on the faces of the students, the instructor should rephrase the question to insure that communication has been effective.

Once the discussion is under way, the instructor should listen attentively to the ideas, experiences, and examples contributed by the students. During the preparation, the instructor should have listed some of the anticipated responses that would indicate the students had an understanding of the concept under discussion. By using "how" and "why" questions, the instructor should be able to guide the discussion toward the objective.

Summary

When it appears the students have discussed the ideas that support this particular part of the lesson, the instructor should summarize what the students have accomplished. An interim summary reinforces learning in relation to a specific objective. In addition, the interim summary also may be used to keep the group on the subject or to serve as a transition to another part of the lesson.

DEMONSTRATION – PERFORMANCE METHOD

The demonstration-performance method of instruction is based on the simple principle that people learn by doing. Physical or mental skills are learned by actually performing these skills after a demonstration of the proper method. A person learns to swim by swimming and to drive by driving. Mental skills, such as speed reading, also are learned by demonstration and performance. Skills requiring the use of tools, machines, and equipment are particularly well suited to this instructional method.

ESSENTIALS OF DEMONSTRATION– PERFORMANCE METHOD

The basic demonstration-performance method of teaching has five essential phases: explanation, demonstration, student performance, instructor supervision, and evaluation. Each phase is discussed in the following paragraphs.

Explanation

In teaching a skill, the instructor must convey the precise actions the students are to perform. In addition to the necessary steps, the instructor may want to describe the end result of these efforts. Before leaving this phase of the method, the instructor should encourage the student to ask questions or, an informal quiz can be given to determine whether the student understands what is to be achieved. For flight maneuvers, this step of the procedure is best accomplished during the preflight briefing. End-level performance criteria should be defined, in addition to the performance limits acceptable at the current stage of training. It is important that end-level performance not be used as a goal at the time the maneuver is introduced, especially if it is a difficult one. At the introduction stage, the instructor might put it this way, "By the time you finish the course, you will be able to hold your altitude within 100 feet during a steep turn but, for your first practice, you will be doing well if you hold it within 200 feet."

Demonstration

In the demonstration phase, the instructor shows the actions necessary to perform the skill. In some cases, the explanation step and demonstration step are accomplished simultaneously. In group instruction, it is sometimes useful for the instructor to explain the procedure while someone else demonstrates. Before leaving the demonstration step, the instructor should attempt to evaluate the students' understanding of the steps involved in the procedure by quizzing or reviewing while watching for facial expressions which might give a clue. When the instructor asks, "Do you understand?" students frequently answer "Yes," even though they do not really comprehend; they are simply reluctant to admit that they did not understand it the first time.

Student Performance and Instructor Supervision

Although student performance and instructor supervision involve separate actions, they are performed simultaneously. For example, the student practices while the instructor watches and offers whatever assistance is required. In group practice, the instructor and whatever assistants are available may circulate among the students as they practice. Students should be allowed to make and correct their own mistakes within reasonable limits, because that is part of the learning process. Learning to correct mistakes is as important as learning to do it right the first time.

Evaluation

In the final phase, the instructor judges student performance. The student is allowed to perform without supervision or instruction and, upon completion of the operation, the instructor comments on the performance, points out errors, offers constructive suggestions for improvement, and points out those things which the student did well. In addition to evaluating the student's performance and readiness to proceed to the next stage, the instructor should evaluate the validity of the instruction.

THE TELLING-AND-DOING METHOD

Another method of flight instruction which actually is based on the demonstration-performance method, also offers several advantages. The method is composed of four separate steps, as illustrated in figure 2-9. Three of the steps involve student activity.

INSTRUCTOR TELLS—INSTRUCTOR DOES

This is the only step in which the student plays a passive role. It combines the explanation and demonstration steps of the previously discussed method. This does not preclude the explanation during the preflight briefing, but only emphasizes that the instructor must make a full explanation concurrently with the demonstration. As the maneuver is performed in flight, control pressures, aircraft attitudes, and any other factors which will assist the student in the performance of the maneuver are explained.

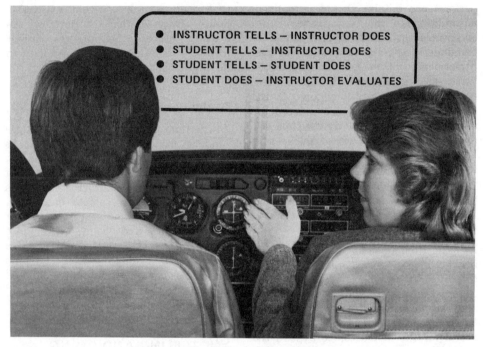

- INSTRUCTOR TELLS — INSTRUCTOR DOES
- STUDENT TELLS — INSTRUCTOR DOES
- STUDENT TELLS — STUDENT DOES
- STUDENT DOES — INSTRUCTOR EVALUATES

Fig. 2-9. The Telling-and-Doing Method

STUDENT TELLS—INSTRUCTOR DOES

This step is the most obvious departure from the demonstration-performance method, and may provide the most significant advantages. In this step, the student plays the role of instructor, telling the instructor what to do and how to do it. Two benefits accrue from this step. First, because of the freedom from the need to concentrate on performance of the maneuver and from concern about its outcome, students are able to organize their thoughts regarding the steps involved and the techniques to be used. In the process of explaining the maneuver as the instructor performs it, their preceptions begin to develop into insights. Mental habits begin to form as they repeat the instructions they have received. Second, with the students doing the talking, the instructor is able to evaluate understanding of the factors involved in performance of the maneuver. If a misunderstanding

exists, it can be corrected before the students become involved in controlling the airplane.

STUDENT TELLS—STUDENT DOES

Students must think about what to do during the performance of a maneuver, until it becomes habitual. In this step, the thinking is done out loud. This focuses concentration on the task to be accomplished, so that total involvement in the maneuver is fostered. All of the students' physical and mental faculties are brought into play. As in the previous step, the instructor is always aware of the students' thought processes, which permits evaluation of the students' level of understanding. It is easy to determine whether an error is induced by a misconception or by a simple lack of motor skill. So, in addition to forcing total concentration on the part of students, this method provides a means for keeping the instructor aware of what students are thinking. Students are not only learning to do something, but are learning how to teach themselves a skill which could be in their favor when they solo.

STUDENT DOES—INSTRUCTOR EVALUATES

When the instructor is satisfied that the student understands the task at hand, it is no longer necessary for the student to talk through the maneuver. This last step, then, is identical to the final step used in the demonstration-performance method. The instructor observes as the student performs, then makes appropriate comments.

INSTRUCTIONAL AIDS

In the classroom, instructors can use instructional aids to improve communication between themselves and their students. Instructional aids may be sight and/or sound devices, which are intended to support, supplement, or reinforce. To plan for the use of aids, it is helpful to understand some of the reasons for using them, and guidelines for their use. It also is helpful to be familiar with different types of aids.

REASONS FOR USE OF INSTRUCTIONAL AIDS

Getting and holding attention are essential to learning. Visual aids which support the topic cause both the seeing and the hearing channels of the mind to work in the same field. These aids make it much easier to hold the student's attention in the classroom.

An important goal of all instruction is for the student to retain as much of the instruction as possible, especially the key points. Many studies have been made to determine the effect of instructional aids on learning retention. While the specific results of the studies vary greatly, all agree that a significant improvement in retention occurs when instruction is supported with meaningful aids.

One problem which plagues instructors is selection of words which have the same meaning for the student as they do for the instructor. This problem is compounded by floods of new technical terms and the differing vocabularies of people involved. Use of visual materials plants the same mental picture in the student's mind that is visualized by the instructor.

Relationships are much simpler to deal with when presented visually. For example, the parts within a mechanical system are easier to relate to each other through the use of schematics or diagrams. Symbols, graphs, and diagrams can also show relationships of location, size, time, frequency, and value. By symbolizing the factors involved, it is even possible to visualize relationships between abstracts.

GUIDELINES FOR USE OF INSTRUCTIONAL AIDS

Any instructional aid must be selected on the basis of its ability to support a specific key point in a lesson. The ideas to be supported with instructional aids are determined from the lesson plan. Aids are often appropriate when long segments of technical descriptions are necessary or when a point is complex and difficult to put into words.

Aids should be as simple as possible and compatible with the learning outcomes to be achieved. Since aids normally are used in conjunction with a verbal presentation, words on the aid should be kept to a minimum. A tendency toward ornateness or unnecessary, distracting artwork should be avoided.

Aids have no value if they cannot be heard or seen. Recordings should be tested for correct volume and quality in the actual classroom environment. Visual aids must be clearly visible to the entire class. All lettering and illustrations must be large enough to be

Fig. 2-10. Types of Instructional Aids

read easily by the students positioned farthest from the aids. If colors are used, they should show clear contrast and be easily visible.

The effectiveness of aids can be improved by proper sequencing. Sequencing can be made relatively simple by using acetate overlays on transparencies, stripping techniques on charts and chalkboards, and imaginative appliques on felt and magnetic boards.

TYPES OF INSTRUCTIONAL AIDS

There are many types of instructional aids available in most schools. In addition to those already available, the imaginative instructor may devise some. The most common aids are illustrated in figure 2-10.

CHALKBOARD

Some instructors find it almost impossible to talk to a class without a piece of chalk in their hand. The chalkboard is one of the most versatile and effective instructional aids if used properly. There are a few practices, however, which are considered fundamental in the use of the chalkboard.

The chalkboard should be kept clean and free of irrelevant material. The equipment, such as chalk, erasers, cleaning cloths, and rulers or other drawing aids, should be kept readily available to avoid interruption of the presentation.

If a chalkboard presentation is planned, it should be organized and practiced in advance. During the presentation, the instructor should write or draw large enough so everyone can see, and step back occasionally to be sure that no one's view is being blocked. The chalkboard should not be overcrowded. A margin should be left around the material, along with ample space between lines of copy. A lively, imaginative chalkboard presentation can be a pleasant learning experience for the students.

PROJECTED MATERIALS

Projected materials include motion pictures and filmstrips, slides, transparencies for overhead projection, and mateials used in the opaque projector. The essential factor for their use is that the content must support the lesson.

Motion pictures can bring many realistic situations into the classroom.

They are usually unsurpassed in their capacity for holding attention. There are many excellent films available on almost any subject. Most can be obtained at a nominal cost on a loan basis, many times for only the cost of postage. Packaged lessons provide an excellent teaching aid, since a common characteristic of these audiovisual presentations is that they are structured rigidly, providing excellent standardization when more than one instructor teaches a course.

MODELS

Models or mockups can be used effectively in explaining operating principles of various types of equipment. They are especially adaptable to small group discussions in which students are encouraged to ask questions. A model is even more effective if it works like the original and can be taken apart and reassembled. As instructonal aids, models are usually more practical than originals, because they are lightweight and easily moved.

Mockups are valuable in teaching such subjects as aircraft systems. They enable the instructor to transmit a clear picture of the position and functions of the components which make up the system.

The effective instructor must keep abreast of developments in instructional aids and in the potential uses for them by reading the professional journals about new instructional materials and by developing a creative imagination. There is always a better way to accomplish a training objective.

PROGRAMMED INSTRUCTION

When programmed instruction is utilized, students speak, write, or make some other response to each increment of instruction. The material offers them immediate feedback (knowledge of how well they are doing) by informing them of the correctness of their response. The successful completion of each of these increments takes students one step closer to the learning objectives.

There are five major characteristics of programmed instruction.

1. Clear specifications of what the students must be able to do after training

2. Careful sequencing of material

3. Presentation of material in steps which challenge students, but do not exceed their ability

4. Active student responses

5. Immediate confirmation of answers

This approach systematically carries students, step by step, to the learning objectives they are to attain. Programmed instruction not only provides students with the information they are to learn, but it also guides them in the way they are to learn.

A number of methods have been used in developing and presenting good programmed instructional materials. One of the most accepted systems is based on rewarding students for accurate performance (reinforcement). In ideal programmed instruction, students cannot make an error. Linear-programmed material is itemized and presented in small steps. Students are prompted, or cued, as necessary, so they almost invariably give a correct response. Materials are carefully designed to offer as much review as needed to insure the degree of retention appropriate to the subject matter, the learning situation, and the needs of the students.

Students respond to a frame (the printed matter for one step or increment of instruction) usually by writing words into spaces provided for that purpose. After completing their response to a frame, students immediately confirm the correctness of their response by comparing it to the programmed answer before continuing to the next frame. In this manner, students progress smoothly, with a continuous satisfaction in being correct.

Linear Method

The sequence of frames shown in figure 2-11, illustrates the linear method of programmed instruction. These sample frames are typical of initial frames, offering obvious cues and sample responses. To the casual observer, this sequence may seem unduly simple. To students who are totally unfamiliar with the subject matter, however, it offers a sort of "learning game" in which they are always a winner. The programmer explains in the introductory materials that students are not being tested, but are being given an opportunity to progress through carefully planned and sequenced material in order to help them learn the new subject matter.

EVALUATION

Instructors determine when they have arrived at their goal through the evaluation process. Evaluation is one of the basic steps in the teaching process carried on throughout each period of instruction. Instructors' evaluation may consist of simple observations of students' rate of comprehension as evidenced by their performance, by the administration of oral or written quizzes, or by formal check flights.

EVALUATION PROCESS

The total evaluation process is the means of determining how well both the instructor and the student are progressing toward certain objectives. Overall objectives are clearly written in the appropriate Practical Test Standards, but the interim objectives for each period of classroom or flight instruction should be determined by the instructor. If the objective is stated in terms of *observable, definable, measurable* behavior which is appropriate to the student's experience, the evaluation process is simple. The student's performance is easily compared to the criteria stated in the objective. If the objective is defined in terms which cannot be observed and measured, however, no two instructors can evaluate the same performance in the same way.

ORAL QUIZZING

Regular and continued evaluation of the student's learning is necessary for

---

1. The three axes about which an airplane can move are:

  _____ (roll), _____

  (pitch), and _____ (yaw).

2. All three axes pass through the _____ of

  _____.

3. The ailerons are movable sections of the outer

  _____ edge of each wing.

4. The ailerons control the movement of the airplane about the

	longitudinal
	lateral
	vertical
	center
	gravity

Fig. 2-11. Linear Programming

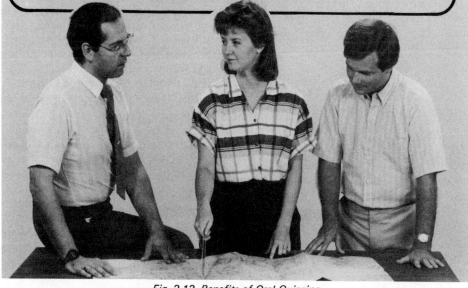

- REVEALS EFFECTIVENESS OF THE INSTRUCTOR'S TRAINING PROCEDURES
- CHECKS STUDENT'S RETENTION OF WHAT HAS BEEN LEARNED
- REVIEWS MATERIAL ALREADY COVERED
- HELPS TO RETAIN STUDENT'S INTEREST AND STIMULATE THINKING
- EMPHASIZES IMPORTANT POINTS OF TRAINING
- IDENTIFIES POINTS WHICH NEED MORE EMPHASIS
- CHECKS STUDENT'S COMPREHENSION OF MATERIAL
- PROMOTES ACTIVE STUDENT PARTICIPATION

Fig. 2-12. Benefits of Oral Quizzing

judging the effectiveness of instruction given and for planning the emphasis and pace of subsequent instruction. The most practical means of evaluation for this purpose is direct or indirect questioning of the student by the instructor. Proper quizzing by the instructor can have a number of desirable results, as shown in figure 2-12.

Effective quizzing requires preparation; good questions are rarely spontaneous. They should solicit specific answers clearly associated with the subject at hand. The questions, "Do you understand?" or "Do you have any questions?" have no place in effective quizzing. Assurance by students that they do understand or have no questions provides no evidence of comprehension. Instead, several specific questions covering important

points in the material presented should be asked, so that evidence of comprehension can be detected.

Other types of questions which should be avoided include catch questions, which may cause students to feel they are engaged in a battle of wits with the instructor; leading questions, which suggest their own answers; irrelevant questions, which destroy the continuity of instruction, and questions which are too broad in scope. For example, the question "What is P-factor?" requires even an instructor to take time to frame an answer. The student's comprehension of the subject can be evaluated better with a series of questions like, "Under what speed and power conditions is P-factor most noticeable?" and "Under those conditions, what must you do to compensate for P-factor?"

Quizzing may be used effectively in several ways. Instructors can question the students, permit the students to ask questions, or present written questions for the students' consideration. The principles of questioning apply to both verbal and written quizzes.

Answers to student's questions must also conform with certain considerations. Instructors must be sure they understand the question before attempting to answer. They should display interest, by words and attitude, in the student's question and frame an answer which is as direct and accurate as possible. As they answer the question, they must watch for clues concerning the student's acceptance and comprehension of the answer because, here again, finishing the answer with, "Do you understand?" will not tell instructors that they have conveyed their message.

Under some circumstances, it may be unwise to introduce the more complicated or advanced considerations necessary to completely answer a student's question. When this is the case, instructors should carefully explain to the student that a good pertinent question has been asked, but that the answer would, at this time, unnecessarily complicate the learning tasks at hand. They should advise the student to ask the question later at the appropriate point in training, if it has not been resolved in the meantime.

Occasionally, students ask questions which instructors cannot answer. In such cases, instructors should freely admit they do not know the answer, but that they will research it. If practicable, they can offer to help students look up answers in the available references. Above all, questions asked by students should not be belittled in any way or treated as anything but a sincere effort to learn. There is no surer way to sever lines of communication than to treat a question lightly or to give the answer less than the consideration it deserves.

DEMONSTRATION OF ABILITY

Demonstrations of piloting ability are basic elements of flight instruction. The instructor determines by quizzing that the student understands the maneuver to be learned, demonstrates the performance, allows the student to practice it under supervision, then evaluates the accomplishment. Evaluation of demonstrated ability during flight instruction must be based on established standards, suitably modified to the student's stage of development.

Demonstrations of piloting ability are important for exactly the same purpose as quizzes. They have additional special significance, however, when directly applied to the qualification of student pilots for solo and cross-country privileges. The stage completion checks conducted in approved flying courses and recommendations also are associated with pilot skill evaluations during flight instruction.

When evaluating demonstrations of piloting ability, as in other instructional processes, it is important for the flight instructor to keep the student informed of progress. Corrections or the explanations of errors in performance should indicate the elements in which the deficiencies are believed to have originated. Then, appropriate corrective measures should be suggested. Correction of students' errors should not include the practice of taking the controls away from them every time they make a mistake. Within reasonable limits, students should be allowed to make mistakes so they can see the result of improper techniques.

At times, the demonstrations of flight maneuvers and operations may be performed correctly by students who do not fully understand the principles involved and the objectives of the exercise. When this is suspected, students should be required to vary their performance of the maneuver slightly, combine it with other operations, or apply the same elements to the performance of other maneuvers. Students who do not understand the principles involved will probably not be able to do this successfully.

EXAMINATIONS AND TESTS

A ground school course should contain written examinations at regular intervals, to evaluate the student's understanding of the training provided and their ability to accept and effectively apply further instruction. For effective appraisal, it is desirable to have many samples of student achievement and growth. Frequent examinations provide students with evidence of their own progress and also inform the instructor of the progress of the entire class and of each individual.

Periodic examinations and a final examination that samples all course areas and objectives, in combination with discussions and interviews, should give an adequate and dependable picture of student achievement. In addition, it allows students to become accustomed to the testing situation, since students are often tense and nervous when they are not familiar with test programs. Experience has shown that students build self-confidence with frequent exposure to tests.

Periodic examinations should be designed to serve as diagnostic tools for both the instructor and the students. A clear picture of weaknesses and strengths emerges through the use of thorough examinations. Both students and instructor then know how to direct their efforts more intelligently during subsequent phases of the course. Competent analyses of the examinations also enables the instructor

to improve the method of instruction and the examinations for future classes.

Many instructors overlook the potential benefit that can result from requiring students to analyze their own examinations during the course. Through such an analysis, students can discover for themselves the nature of their strengths and weaknesses. They may record their errors and indicate test or workbook material which requires further study. Used in this way, periodic examinations become effective teaching devices in themselves.

Although final examinations can be used to determine final standing, that should not be their sole purpose. Like periodic examinations, end-of-course examinations should be designed to yield useful information about the instruction. Sound analyses of final examination results can indicate the degree to which the overall objectives of instruction have been achieved. The examinations also may predict later job successes, and indicate trends in teaching effectiveness, and the need for revisions in teaching techniques. For additional information on how instructors can write their own tests, the applicant should refer to *AC 60-14 — Aviation Instructor's Handbook.*

APPLYING APPROPRIATE STANDARDS

The criteria used for final evaluation and for stage tests and checks are prescribed in the school's training syllabus and the appropriate flight test guides. The standards which must be applied by instructors in their day-to-day instruction, however, usually must be established or at least administered personally.

The criteria used in the course of instruction cannot be standardized because no two people think, react, and progress in an identical manner. One student may respond well to instruction in stall recognition and perform acceptably after a minimum amount of practice, while the next may experience difficulty and require prolonged assistance and practice to achieve the same level. This one element of performance may not be at all indicative of the student's relative piloting ability. For example, the situation may be reversed in the case of coordination in turns.

Some students may be slow but steady learners, while others may grasp new learning tasks quickly, but make many careless mistakes. Professional flight instructors must learn to size up their students, become familiar with their thinking and aptitudes, and apply standards of progress appropriate to each individual. Attempts to make students conform to an arbitrary general standard may actually retard their overall progress.

The application of instructional and progress standards tailored to each student must not lead to the modification of performance standards for pilots prepared for certification. All pilots trained and endorsed for solo privileges or recommended for certification must meet the appropriate knowledge, skill and proficiency standards. Only their rate of training and the method of individualizing the instructional presentations should be modified.

THE INSTRUCTOR AS A CRITIC

Among the many skills instructors possess, none is more important than the ability to analyze, appraise, and judge the performance of their students. Students quite naturally see their instructors as critics and look to them for guidance, analysis, appraisal, suggestions for improvement, and encouragement. Criticism in the classroom is recurring and constant,

but sometimes it is formalized, put into a structure, scheduled as part of a class period, and termed a critique.

A classroom critique may be either oral or written. It should come immediately after an individual or group performance, such as an examination, while the details of the performance are easy to recall. Instructors may critique any activity which students perform or practice to improve their skill, proficiency, and learning. It may be conducted in private or before the entire class. A critique presented before the entire class can benefit every student in the classroom, as well as the one who performed the assignment.

Two common misconceptions about the critique should be corrected at the outset. First, a critique is not a step in the grading process, and should be considered apart from it. It is a step in the learning process, and should occur before formal measurement takes place. Second, a critique is not necessarily negative in content. It considers both the good and the bad; it can be as many sided and varied in content as the performance.

A critique should improve the students' future performance and should give them something to work with; some direction and guidance to raise their level of performance. However, unless students understand the purpose of the critique, they cannot accept the criticism offered and little improvement can be expected.

EFFECTIVE CRITIQUES

An effective critique has several major characteristics. It should be objective and focused on the students and their performance, without reflecting the personal opinions of the instructor. It should be flexible; the tone, technique, and content must fit the situation and the students. A critique must have acceptability. Before students willingly accept their instructors' criticism, they must first accept the instructors themselves. They must have confidence in the instructors' qualifications, teaching ability, sincerity, and competence. A critique also must be comprehensive. This does not necessarily mean that it must be long or treat every aspect of the subject in detail. Instead, instructors must decide on the benefits to be derived from the discussion of a few major points or a number of minor points.

An effective critique covers strengths as well as weaknesses. To dwell on excellence and neglect the portion that should be improved is a disservice to students. A critique is pointless unless students profit from it. Praise for the sake of praise is of no value if students are not taught how to capitalize on it. By the same token, negative criticism that does not point toward improvement should be omitted from a critique. Unless a critique follows some pattern or organization, a series of otherwise valid comments may lose their impact. Almost any pattern is acceptable so long as it makes sense to the students, as well as to the instructor.

An effective critique recognizes the instructor's thoughtfulness toward the students' needs. The inherent dignity and importance of the individual should never be minimized. Ridicule, anger, or fun at the expense of students has no place in a critique. The instructor's comments and recommendations should be specific. At the conclusion of a critique students should have no doubt about what they did well, what they did poorly and, most important, specifically how they can improve themselves.

The critique of a student's performance is always the instructor's responsibility, and it should never be delegated. Interest and variety can be added, however, by trying different methods. One way is for the instructor to lead a group discussion in which members of the class are invited to offer criticism. Certain ground rules must be established in advance to prevent the critique from deteriorating into a free-for-all.

There are various types of critiques that can be conducted by the students. For example, the instructor might ask a student to lead the critique. Because of the inexperience of the participants, student-led critiques may not be as efficient as those led by the instructor but, by stimulating student interest and learning, they can be effective. The class may be organized into small groups, with each one assigned an area to critique, or the instructor might assign a single student to present the entire critique. Students can be asked to criticize their own performance. This technique can be very useful in flight training. In criticizing themselves, students may reveal a misunderstanding about one or more elements of the operation.

Whatever the methods employed, the instructor must not leave controversial issues unresolved or erroneous impressions uncorrected. If the students participate in the critique, the instructor must make allowances for their relative inexperience. However, time should be reserved at the end of the critique to cover those areas that might have been omitted or not emphasized sufficiently.

CRITIQUING RULES

There are certain ground rules which must be observed for the critique to accomplish its intended purpose.

1. The critique should not be extended beyond its scheduled time, except in rare and unusual instances.

2. The instructor should avoid trying to cover too much material. A few well-made points may be more

beneficial than a large number of points not developed adequately.

3. A critique should not be stretched beyond its normal boundaries merely to fill a class period. A point of diminishing returns can be reached quickly.

4. Time should be allowed for a summary of the critique to reemphasize the most important points.

5. Dogmatic or absolute statements should be avoided, since most rules have exceptions.

6. Controversies with the class should be avoided.

7. Instructors should never allow themselves to be maneuvered into the position of defending their criticism. If the criticism is honest, objective, constructive, and comprehensive, no defense should be necessary.

8. If part of the critique is written, it should be consistent with the oral portion.

SECTION C—PLANNING AND ORGANIZING

A successful teaching endeavor can be attributed to extensive planning and well thought-out organization. Much of the basic planning for flight and ground instruction is provided by requirements of the Federal Aviation Regulations; approved school curriculums; and the various texts, manuals, and training courses available. This section reviews the planning required of the professional flight or ground instructor as it relates to three topics, *course of training, training syllabus,* and *lesson plan.*

COURSE OF TRAINING

Before instruction can begin, a determination of objectives and standard is necessary. In the case of a pilot training course, the overall objective is simply stated and the minimum standards are provided by Federal Aviation Regulations and Practical Test Standards. For example, a course objective and completion standard for a private pilot training course should be similar to the one shown in figure 2-13.

The general overall objective of any pilot training course is designed to qualify the student to be a competent, efficient, safe pilot while operating specific aircraft types under stated conditions. The criteria for determining whether the training has been adequate are the successful completion of written and flight tests required by the Federal Aviation Regulations for the issuance of pilot certificates.

The conscientious instructor, however, does not limit training objectives to merely meeting the minimum published requirements for a pilot certificate. Additional objectives include, training each student to service an airplane properly, to maneuver and operate it accurately within its limitations, and to analyze and make prompt decisions with respect to its safe operation. This is only a partial list of general objectives, but it illustrates the major planning which is the basis of any training endeavor.

BUILDING BLOCK PRINCIPLE OF LEARNING

It is difficult for anyone to learn a complex skill without following a step-by-step sequence. The classic example of this principle is the baby learning to crawl before walking. To carry this principle further, the child must master walking before attempting to run. It is easy to see that, if one phase is omitted, the objective will be more difficult, if not impossible to achieve.

OBJECTIVE

The student will obtain the knowledge, skill and aeronautical experience necessary to meet the requirements for a private pilot certificate with an airplane category rating and a single-engine land class rating.

COMPLETION STANDARD

The student must demonstrate through written tests, flight test, and show through appropriate records that the knowledge, skill and experience requirements necessary to obtain a private pilot certificate with an airplane category rating and a single-engine land class rating have been met.

Fig. 2-13. Course Objective and Completion Standard

Blocks of learning should be organized when planning a course of flight training so that subject areas are reasonably compatible and presented in logical sequence. For example, a block of ground instruction dealing with weather theory and weather reports and forecasts fits logically into the training sequence before the first dual cross country. It would be impractical to cover cross-country planning in a ground training lesson after the cross country sequence of flight training. Therefore, it is important that, after blocks of learning have been identified for both ground and flight training, they be integrated in a logical manner. The ground training should always proceed in advance of related flight training so that maximum benefit is derived from each fight lesson.

MEANINGFUL REPETITION

Another valuable criterion to apply to a course of training is the amount of meaningful repetition that it incorporates. Meaningful repetition simply means that each necessary bit of knowledge, concept, or skill is presented to the student several times throughout the instructional program. Each repetition is varied enough from the previous presentation to constantly present the student with a new slant, a different perspective and, therefore, a challenge. This usually is accomplished with the use of several distinctly different learning tools. These might include audiovisuals, a textbook, a workbook, class discussions, exams, and the skills learned in the airplane. It is obvious that a course of training which provides this type of repetition has several advantages for the student.

TRAINING SYLLABUS

The form of the syllabus may vary, but it is always in the form of an abstract or digest of the course of training. It consists of the blocks of learning to be completed in the most efficient manner. Advisory Circular 141-1 defines a

training syllabus as *a step-by-step (building block) progression of learning with provisions for regular review and evaluation at prescribed stages of learning. The syllabus defines the unit of training; states, by objective, what the student is expected to accomplish during the unit of training; shows an organized plan for instruction (building block—from the simple to the complex); and dictates the evaluation process for either the unit or stages of learning.* The length of each lesson and stage in the syllabus should be established on the basis of units of learning, not an arbitrary time allotment.

Effective use of the syllabus requires the instructor to refer to it throughout the entire course of training. However, it should not be adhered to so stringently that it becomes inflexible or unchangeable. The various rates of learning encountered will require some repetition and alteration to fit particular individuals. The syllabus also should be flexible enough so it can be adapted to weather conditions, aircraft availability, and scheduling changes without disrupting the teaching process or completely suspending training.

PLANNING AND ORGANIZING THE GROUND TRAINING LESSON

The particular subject covered during a ground training lesson is only one factor which will affect the presentation of the lesson. Other factors include the availability of teaching aids such as films or audiovisual programs, models, mockups, and the training facility itself. Probably the most important consideration is whether the presentation will be made to a class or an individual.

INDIVIDUAL INSTRUCTON

Providing ground instruction to one individual at a time can have several advantages; for example, scheduling is one of the benefits of this training method. A ground training lesson can be placed just prior to a related training flight or series of flights. In addition, the instructor can concentrate on the areas of the lesson which the student identifies as most difficult. In this way, each lesson can be tailored to the individual and efficient use of the training time results. Finally, the student can progress at his or her own rate, which usually makes for more meaningful training, better student/instructor relations, and allows scheduling around the day-to-day commitments outside of learning to fly.

CLASSROOM INSTRUCTION

Many operators and instructors prefer to conduct ground training in a formal classroom environment. This method also has several advantages which include efficient use of the instructor's time and lively discussions between the class members and the instructor.

Probably the greatest benefit is that the students learn from each other. In addition, classroom instruction tends to introduce a healthy competitive element to the teaching and learning process.

Admittedly, classroom instruction has some obvious drawbacks. For example, the classroom method requires that enough students be enrolled at a particular time to start the class, which means that some students may be held up in the flight training segment because of the ground training schedule. There also is the problem of achieving good integration of the ground and flight training segments. Even though the ground lesson may have many participants, the teaching environment can be altered to make the experience as pleasing as possible, through the use of a well lighted, spacious classroom with plenty of chalkboard space and all available teaching aids.

STRUCTURE OF THE GROUND TRAINING LESSON

Ground training lessons in a training syllabus generally are organized as follows: objective, lesson content, completion standard, and study assignment. The elements of the lesson should be designed so they are applicable to individual or classroom instruction. Figure 2-14 illustrates a typical ground training lesson. This illustration should be referred to during the following discussions of each element.

OBJECTIVE

The ground training lesson objective states, in simple terms, what is to be learned during a particular unit of training. The instructor should reiterate the lesson objective during the introduction to emphasize to the student or class what will be accomplished. The objective is also closely related to the completion standard which will be discussed later.

CONTENT

The content of the lesson should be listed in the logical order of introduction. In most commercially prepared training courses, the content will follow the order of presentation used for the training materials. For example, a given lesson may cover the material in a specific chapter of the course textbook. However, instructors who prepare their own syllabi may organize the lesson presentation to directly complement the flight training program. In any event, the content of each lesson must blend in logically with other lessons to form the total ground training course.

COMPLETION STANDARD

Completion standards are of no use to the student or instructor without a method of evaluation to reveal whether or not they have been met. That is to

STAGE II
GROUND LESSON 6
TEXT REFERENCE:
Private Pilot Manual—Chapter 5, "Interpreting Weather Data"

VIDEO PRESENTATION:
Private Pilot Course—Volume 5, "Interpreting Weather Data"

RECOMMENDED SEQUENCE:
1. Lesson Introduction and Video Presentation
2. Class Discussion

LESSON OBJECTIVE:
During this lesson, the student will learn how to procure and interpret weather reports, forecasts, and charts. In addition, the student will become familiar with the various sources of weather information.

CONTENT:
Section A—"Printed Reports and Forecasts"
___ Surface Aviation Weather Reports
___ Radar Weather Reports
___ Terminal Forecasts
___ Area Forecasts
___ Winds and Temperatures Aloft Forecasts
___ Severe Weather Reports and Forecasts
Section B—"Graphic Weather Products"
___ Surface Analysis Chart
___ Weather Depiction Chart
___ Radar Summary Chart
___ Low-Level Significant Weather Prog
Section C—"Sources of Weather Information"
___ Preflight Weather Briefings
___ Supplemental Weather Sources
___ In-Flight Weather Services
___ In-Flight Weather Advisories

COMPLETION STANDARDS:
The student will complete *Private Pilot Exercises 5A, 5B,* and *5C* with a minimum passing score of 80%, and the instructor will review each incorrect response to ensure complete understanding before the student progresses to Ground Lesson 7.

STUDY ASSIGNMENT:
FAR Booklet—Private Pilot FARs

Fig. 2-14. The Ground Training Lesson

say, completion standards should be stated in terms of measurable units of human accomplishment. Normally, the lesson evaluation is in the form of workbook exercises or a quiz. This method of evaluation allows the instructor to judge the effectiveness of the instruction as well as the student's performance. If several individuals score low on a phase of instruction, the instructor should question the accuracy of the source material or the

effectiveness of the teaching process. Oral quizzing during ground training also can be used effectively to supplement written evaluations. In an individual training situation, the oral quiz is especially effective because every question is directed at one person.

STUDY ASSIGNMENT

At the completion of each ground lesson, a study assignment should be made so the student can prepare for the next lesson. If emphasis on a particular subject is planned for the next lesson, it should be announced at the time of assignment.

Energetic students will consistently read ahead of the assignment for the rest of the class. Questions about material not yet covered should be answered before or after the scheduled ground session. Answering questions of this type during the class will only confuse those who have only prepared for the assigned lesson.

AUDIOVISUALS

Flight and ground training courses supplemented by good audiovisual presentations now are becoming increasingly popular. Most audiovisual programs are designed to motivate the student to make an individual effort to study and learn the training material and, thereby, attain the desired certificate or rating. Based on this premise, the audiovisual program can be used to introduce new subjects and to reinforce the concepts presented in other materials. Audiovisuals also can be used as the basis for class discussions or lectures.

If the audiovisual program precedes the appropriate lecture, it introduces the subject and provides the class with a broad overview of lesson content. A quality audiovisual presentation will help to maintain a high level of interest during the entire lesson.

TIME ALLOTMENT

Each ground lesson should be planned so it can be completed in a predetermined unit of time. If the material cannot be covered in that length of time, a separate lesson should be scheduled for its completion. The maximum length of a lesson should be limited to about two hours. Interest will deteriorate rapidly if longer lessons are utilized.

PLANNING AND ORGANIZING THE FLIGHT TRAINING LESSON

Planning and organizing a lesson for an individual is, in theory, the same as for a large group of people. The only significant difference is in the flexibility of the lesson. It is much easier for an instructor to identify areas which require greater emphasis when the objective of the lesson is directed to one person. This early identification of problems allows the instructor to immediately alter the lesson content, or even repeat some elements of an earlier lesson.

Since every student is a unique individual, a slightly different teaching technique may be necessary to achieve the most beneficial results. As an example, the overconfident student may require strict discipline to instill safety. On the other hand, strict discipline might not be required for a highly responsible, self-disciplined student.

The flight lesson also is an ideal teaching situation from the standpoint of student/instructor ratio. With the one student/one teacher relationship, every student obtains individually modified and adapted instruction. This provides the instructor with an ideal opportunity to provide a very effective learning experience for the student.

STRUCTURE OF THE FLIGHT TRAINING LESSON

Flight training lessons are very similar in structure to ground lessons in that they both contain sections that describe the lesson objective, content, completion standards, and a study assignment. The differences between the two are in the application of the lesson itself. The preflight and postflight briefings have been added to the lesson content, and the completion standards must be evaluated in a different manner. Figure 2-15 is an example of a typical flight lesson.

**STAGE II
FLIGHT 9
DUAL AND SOLO—LOCAL**

RECOMMENDED SEQUENCE:
1. Preflight Orientation: Briefing—Solo Flight
2. Flight
3. Postflight Evaluation

LESSON OBJECTIVE:
During the dual portion of this lesson, the instructor will review takeoff and landing procedures to check the student's readiness for solo flight. During the second portion of the lesson, the student will fly the first supervised solo flight in the local traffic pattern.

CONTENT:
Lesson Review
___ Engine Starting
___ Radio Communications
___ Normal and/or Crosswind Taxi
___ Pretakeoff Check
___ Normal Takeoffs
___ Traffic Pattern Operations
___ Go-Around From a Rejected Landing
___ Normal Landings

Lesson Introduction
Supervised Solo
___ Radio Communications
___ Taxi
___ Pretakeoff Check
___ Normal Takeoffs and Climbs (3)
___ Traffic Pattern Operations
___ Normal Approaches and Landings (3)
___ Postflight Procedures

COMPLETION STANDARDS:
The student will complete the supervised solo as directed by the instructor. The student will adhere to established traffic pattern procedures and demonstrate that solo flight in the traffic pattern can be accomplished safely.

*LESSON ASSIGNMENT:
Ground Lesson 7—Page 2-30

Fig. 2-15. The Flight Training Lesson

OBJECTIVE

The objective of the lesson describes what should be learned upon the completion of a flight lesson. The objective of a flight lesson should be developed so that evaluation of performance during the lesson is possible.

CONTENT

The preflight orientation should be considered an integral part of the lesson content. For private pilot applicants, a preflight briefing should be provided for each dual and solo flight. Commercial applicants should receive one before each dual flight and before selected solo flights. The preflight briefing provides an explanation of the subject matter to be covered during the lesson. It is important that the instructor define new or unfamiliar terms and explain the objectives of the lesson.

A review of the content of the previous lesson is necessary to reinforce principles that will apply to material introduced in the current lesson. If a basic principle has been forgotten, the successful completion of the lesson probably will be in doubt. After a thorough review has been completed, new principles or techniques should be introduced. If the introduction of new material immediately follows the review, the student will be able to more readily associate related principles.

At the end of each flight session the instructor should conduct a postflight briefing. Private pilot applicants should receive a postflight briefing after each dual and solo flight. Commercial applicants should be provided with a briefing after each dual flight and after selected solo flights. This provides an opportunity for the instructor to critique the student's performance and make necessary recommendations for improvement. It is a valuable instructional technique because it increases retention and, to some degree, prepares the student for the next lesson.

COMPLETION STANDARDS

The completion standards for the flight lesson must be developed so they can be judged during the course of the lesson. The measurable unit of human behavior, in this case, is the skill and accuracy demonstrated during the performance of a maneuver. The completion standards usually are stated in terms of airspeed, altitude, and heading tolerances.

As the flight training progresses, the completion standards should become progressively higher. Each lesson then has completion standards that are appropriate for the stage of training being conducted. If the level of performance desired from a student is challenging, yet attainable, a higher degree of interest will be maintained throughout the entire course of training.

STUDY ASSIGNMENT

At the end of each lesson, a study assignment should be made. This technique prepares the student for the next flight lesson, and allows the opportunity for some preparation. Considerable time can be saved during the introduction of new concepts if the student is properly acquainted with what to expect during the next flight.

LESSON PLAN

Advisory Circular 141-1 defines lesson plan as "the instructor's plan for teaching a unit of learning. It is the basic method for an orderly flow of information to a student based on the student's way of learning."

It can be said that instruction is adequately planned only when the instructor has a lesson plan for each period or unit of instruction. As stated in the previous discussions, the lesson plan is the culmination and direct application of general and specific planning which must be the basis of all effective instruction. Teaching success depends more on lesson planning than it does on presentation, personality, flying ability, or experience. Teaching is somewhat like a battle, in that effort, strength, and sincerity will not win if the strategy of its conduct is faulty. By the same token, the finest workmanship and materials will not build a good airplane if the basic design is faulty.

An experienced flight instructor who has trained many students is able to construct an effective lesson plan for a routine period of instruction instinctively, or at least without committing it to writing. However, an instructor who has been through the course only a few times, or an experienced instructor who must modify teaching procedures to effect special emphasis, should always prepare a written lesson plan. Lesson plans can be brief, topical in nature, and without a prescribed format. It is prepared for the instructor's own benefit, and should be done in the form that is most useful to that individual.

The lesson plan may be more or less detailed and may include the special or associated considerations which should be covered during an instruction period. A lesson plan prepared for one student rarely is appropriate to another without some modification.

The lesson plan can be carried by the instructor as a checklist to assist in the administration of the lesson, or it can be studied until the instructor is confident that no diversions to the planned procedure will occur.

Once a lesson has begun, the instructor should not allow the application of the lesson plan to be diverted to other subjects and procedures. This does not mean that the planned instruction should not be modified by circumstances or by the discovery of pertinent deficiencies in the student's knowledge or in performance of elements essential to its effective completion. In some cases, the whole lesson may have to be abandoned in favor of a review of knowledge and operations covered previously.

To facilitate this, each lesson should begin with a brief review of elements covered during previous lessons and any practice necessary to bring the student's performance up to the proficiency assumed for the start of the present lesson. If this review grows to unanticipated proportions, or necessitates the abandonment or significant revision of the lesson plan, the instructor must be prepared to mentally construct a new lesson plan to guide the remainder of the instruction period.

The mechanics of constructing a lesson plan for each period of instruction may seem unduly burdensome. However, the conscientious development and use of lesson plans is the most effective means of developing orderly and effective teaching habits. The procedure soon becomes habitual and each segment falls into place with little effort on the part of the experienced instructor.

The use of a standard, or prepared, lesson plan for all students rarely is effective because each student requires a slightly different approach. Assistance from an experienced flight instructor in preparing lesson plans is often helpful for a new instructor, but a lesson plan prepared by someone else is not as helpful to an instructor as one which is personally prepared.

Flight 6A   Student: John Smith

Dual-Local
(7 to 10 Knot Crosswind Conditions Required)
Sequence:
1. Preflight Orientation
2. Flight
3. Postflight Evaluation

Lesson Objective:
   During the lesson, the student will review
crosswind landing techniques in actual
crosswind conditions and attempt to increase
understanding and proficiency during their
execution. The principle of a stabilized
landing approach will be emphasized.

Lesson Review
1. Slips
2. Crosswind Landings

Completion Standards:
   The student will demonstrate an understanding of
how the slip is used to perform crosswind
landings. In addition the student will demonstrate safe crosswind landings in light
crosswind conditions.

Notes: Emphasize that the runway, airplane
path, and longitudinal axis of airplane must
be aligned at touchdown. Have the student establish a slip on "high final approach" rather
than crabbing and establishing slip just prior
to touchdown. This should allow the student
to concentrate on keeping the upwind wing low while
maintaining runway alignment during the flare.

Fig. 2-16. Lesson Plan

Typically, an instructor will adapt a lesson plan from a training syllabus for a specific student. It is a rare occasion when a student can complete a program of instruction without some modification in the sequencing of training to solve a particular problem.

The following example is a hypothetical situation designed to illustrate how a lesson plan can be developed to supplement a lesson in the training syllabus. Assume a student who is experiencing difficulty with crosswind landings, is unable to meet the completion standards of the flight lesson. The instructor can elect to develop a specific lesson on this subject. Figure 2-16 illustrates a sample supplemental lesson and includes notes concerning the conduct of the special lesson.

chapter 3

INTRODUCTION

Flight instructors have gradually received increased recognition from the aviation industry in general. In addition, fixed base operators, realizing the importance of a flight instructor's role, have invested more of their resources to improve the training segment of their operations. For the first time, many individuals look upon flight and ground instruction as a professional and permanent occupational endeavor, rather than a stepping stone to some other career in aviation. The Federal Aviation Regulations as well as the practical test standards have placed increased reliance on instructors to provide overall supervision of the training and certification of pilots.

Along with the expanding role of the instructor goes an increased responsibility to students and the general flying public. To meet these reponsibilities, professionalism has become a necessary prerequisite for the successful instructor. This chapter defines some of the qualities of the professional instructor and provides other information of practical value to the beginning as well as the experienced instructor. It also contains an analysis of training requirements for the various pilot certificates and ratings and shows the full range of instructor endorsements. In addition, the portion of FAR Part 61 pertaining to the certification of flight instructors is presented in the last section for the benefit of the flight instructor applicant.

SECTION A—AUTHORIZED INSTRUCTION AND ENDORSEMENTS

Obviously flight instructors must be fully qualified pilots. However, their qualifications must go far beyond those required for pilot certification if they expect to receive recognition as professional flight instructors.

PROFESSIONALISM

Although the word "professionalism" is used often, it is rarely defined, since a single definition will not encompass all of the qualifications of a professional.

Some of the major qualities of professionalism are shown in figure 3-1. Attempts to operate as a flight instructor without all of the qualities listed can result in poor performance and inadequately prepared students. Professional flight instructors constantly emphasize preparation and performance, and they take specific steps to effectively foster learning. They clearly define goals and objectives, devise a plan of action, create a positive instructor-student relationship, present information and guidance effectively, and transfer responsibility to the students as they learn.

All of these factors combine to describe the total performance of an instructor. While these distinct factors may not be apparent to the student, disregarding any one of them leads to a difficult or inefficient learning experience. Their proper application can result in a learning experience which is effective and satisfying for the student.

RESPONSIBILITIES TO STUDENT PILOTS

The basic function of the flight instructor is to provide the flight and ground instruction appropriate to the student's needs. Competent flight instruction must be appropriate to the individual student and the circumstances under which it is given. The slow student requires special handling to promote confidence and avoid discouragement. Assigning "subgoals" which are easily achievable is a useful technique.

In contrast, a good student may exhibit overconfidence. This tendency can be corrected by increasing the standard of performance required for each lesson. Instructors must remember that students generally do not ask for the information or guidance needed, and they normally do not understand how the learning tasks assigned apply to overall objectives. It is the responsibility of the instructor to see that each period of

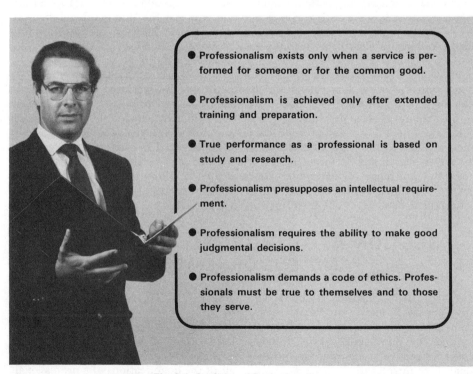

- ● Professionalism exists only when a service is performed for someone or for the common good.

- ● Professionalism is achieved only after extended training and preparation.

- ● True performance as a professional is based on study and research.

- ● Professionalism presupposes an intellectual requirement.

- ● Professionalism requires the ability to make good judgmental decisions.

- ● Professionalism demands a code of ethics. Professionals must be true to themselves and to those they serve.

Fig. 3-1. Qualities of Professionalism

learning provides another block of instruction in its proper position in the total structure, and to help the student see the relationship.

DEMANDING AN ADEQUATE STANDARD OF PERFORMANCE

Flight instructors must constantly evaluate their own effectiveness and the standard of learning and performance achieved by their students. The desire to maintain pleasant personal relationships with students must not cause the acceptance of a slow rate of learning or a low level of flight performance. Reasonable standards when strictly enforced are never resented by an earnest student. Flight instructors actually fail to provide some of the instruction their students are paying for if they permit them to get by with substandard performance or without learning some items of knowledge pertinent to safe piloting. An even more important effect of this improper training is that such deficiencies may later allow hazardous inadequacies in the student's performance as a pilot.

STUDENT PILOT SUPERVISION

Flight instructors have a moral obligation to provide guidance in respect to student's solo operations. If unsatisfactory performance is observed, it is the instructor's responsibility to try to correct it by the most reasonable and effective means.

AUTHORIZED INSTRUCTION

Each flight instructor is granted certain privileges by the Federal Aviation Administration upon certification. These privileges include the authorization to give flight instruction under specified conditions, provide necessary endorsements, and recommend applicants for pilot certification tests. The following presentation describes these conditions.

REGULATORY BASIS

The requirements for the issuance of pilot and flight instructor certificates and ratings are prescribed by FAR Part 61. In addition, FAR Part 61 states the conditions under which those certificates and ratings are necessary, as well as the associated privileges and limitations. FAR Part 141 governs the certification of *approved pilot schools* and sets forth the rules under which they operate. Since these pilot schools must conduct flight and ground training under specific guidelines and meet rigid operational requirements, graduates of these schools are permitted certification with less total flight experience than that specified in FAR Part 61. A familiar example is the total flight time requirements for private pilot certification. Under FAR Part 61, 40 hours is the minimum; whereas, a graduate of an approved school is required to have only 35 hours.

AERONAUTICAL EXPERIENCE

Each certificate or rating requires a different level of aeronautical experience. The knowledge and experience required for each certificate and rating is defined explicitly in FAR Part 61. The following is a summary of the aeronautical experience required for each certificate or rating and, where appropriate, a comparison of the training requirements of FAR Part 61 and FAR Part 141.

PRIVATE PILOT (AIRPLANE)

Total (FAR Part 61)	40 hours
Dual	20 hours
Cross country	3 hours
Night (including 10 takeoffs and landings)	3 hours
Flight test preparation	3 hours
Solo	*20 hours
In airplanes	10 hours
Cross country	10 hours

*Includes three solo takeoffs and landings to a full stop at an airport with an operating control tower.

Total (FAR Part 141)	35 hours
Ground trainer maximum	5 hours
Dual	20 hours
Solo	15 hours
Cross country	10 hours

COMMERCIAL PILOT (AIRPLANE)

Total (FAR Part 61)   250 hours
  Ground trainer maximum 50 hours
  Powered aircraft   100 hours
    In airplanes   50 hours
  Dual   50 hours
    Complex airplane   10 hours
    Instrument instruction 10 hours
      In airplanes   5 hours
    Flight test preparation 10 hours
  Pilot in command   100 hours
    In airplanes   50 hours
    Cross country   50 hours
    Night   *5 hours
*Includes 10 takeoffs and landings.

Total (FAR Part 141)   190 hours
  Ground trainer maximum 40 hours
  Dual   75 hours
    Ground trainer
      maximum   20 hours
    Complex airplane   10 hours
    Night flight including cross-
      country flight   no minimum
                            stated
  Solo   100 hours
    Cross country   40 hours
    Complex airplane   *5 hours
    Night   *5 hours
*Includes 10 takeoffs and 10 landings to a full stop.

AIRLINE TRANSPORT PILOT (AIRPLANE)

Total (FAR Part 61)   1,500 hours
  Cross country   500 hours
  Night   100 hours
  Instrument (actual or
    simulated)   75 hours
    In flight   50 hours
  Pilot in command
    or copilot   250 hours
    Cross country   100 hours
    Night   25 hours

Total (FAR Part 141) Same as Part 61
                    experience
                    requirements
  Dual   25 hours
  Instrument instruction
    minimum   15 hours

INSTRUMENT RATING (AIRPLANE)

An applicant for an instrument rating must hold at least a current private pilot certificate with an appropriate aircraft rating.

Total (FAR Part 61)   125 hours
  PIC cross country   *50 hours
  Instrument (simulated or
    actual)   40 hours
    Ground trainer instruction
      maximum   20 hours
    Dual (instrument
      instruction)   15 hours
      In airplanes   5 hours
Total (FAR Part 141)   125 hours
  PIC cross country   *50 hours
  Dual (instrument instruction) 35 hours
    Ground trainer maximum 15 hours
*In powered aircraft with other than a student pilot certificate.

ADDITIONAL AIRCRAFT RATING

Additional aircraft ratings apply to situations where an individual possesses a pilot certificate and desires to add additional aircraft category, class, or type privileges. The following ratings can be placed on pilot certificates initially or may be acquired later as an additional rating.

Aircraft category ratings
  Airplane
  Rotorcraft
  Glider
  Lighter-than-air
Airplane class ratings
  Single-engine land
  Multi-engine land
  Single-engine sea
  Multi-engine sea
Rotorcraft class ratings
  Helicopter
  Gyroplane
Lighter-than-air class ratings
  Airship
  Free balloon

Aircraft type ratings are listed in Advisory Circular 61-1, entitled "Aircraft Type Ratings."

An applicant for a *category rating* to be added on a pilot certificate must meet the requirements of FAR Part 61 for the issue of the pilot certificate appropriate to the privileges for which the category rating is sought. However, the holder of a category rating for powered aircraft is not required to take a written test for the addition of a category rating to a pilot certificate.

There is no established minimum amount of flight time necessary in order to add an additional *aircraft class rating* to a pilot certificate. FAR Part 61 requires instruction be received appropriate to the desired rating, and that a flight instructor recommendation be obtained. Of course the appropriate flight test also must be successfully completed.

FAR Part 141 differs from FAR Part 61 only in the training required for a type rating. An approved school must provide at least 10 hours of flight instruction for a type rating course.

FLIGHT INSTRUCTOR CERTIFICATE

FAR Part 61 requires that applicants for flight instructor certificates hold either a commercial pilot certificate with an instrument rating or an airline transport pilot certificate. Under FAR Part 61, an additional flight instructor rating requires that the applicant have at least 15 hours as pilot in command in the category and class of aircraft appropriate to the rating desired and pass the written and practical tests.

Flight instructor certification under FAR Part 141 requires that the applicant hold a commercial pilot certificate prior to enrollment. If an airplane instructor rating is desired, the applicant must hold an instrument rating.

The approved school certification course must include at least 25 hours of instructor training in addition to the required ground training. This instructor training involves 10 hours of flight instruction in the analysis and performance of flight training maneuvers, five hours of practice ground instruction, and 10 hours of practice flight instruction (with the instructor in the aircraft). To add a rating to a flight instructor certificate, the instructor training must include at least 20 hours (10 hours of flight maneuvers analysis and 10 hours of practice flight instruction) in addition to the ground training.

AERONAUTICAL KNOWLEDGE

FAR Part 61 does not specify a given number of hours of ground instruction as a requirement for a pilot or flight instructor certificate or rating. It does, however, specify the *general subjects* in which an applicant must have either received ground instruction or completed a home study course. The FAR Part 61 references appropriate to the aeronautical knowledge requirements for the various pilot and flight instructor certificates and ratings are shown in the following list.

1. Student Pilot 61.87 (b)
2. Private Pilot 61.105
3. Instrument Rating 61.65 (b)
4. Commercial Pilot 61.125
5. Airline Transport Pilot 61.153
6. Flight Instructor 61.185
7. Additional Instructor Rating 61.191 (c)

EVIDENCE OF GROUND INSTRUCTION OR HOME STUDY COURSES

Under FAR Part 61, an applicant for a pilot or flight instructor written test is required to show satisfactory completion of the ground instruction or home study course required for the certificate or rating sought. A home study course for the purposes of Part 61 is a course of study in those aeronautical subject areas specified by the applicable regulation and organized by a pilot school, publisher, flight or ground instructor, or by the student. The applicant may develop a home study course from material described in the appropriate FAA question book. Any one of the following may be accepted as evidence of meeting this requirement:

1. A certificate of graduation from a pilot training course, appropriate to the certificate or rating sought conducted by an FAA certificated pilot school, or a statement of accomplishment from the school certifying the satisfactory completion of the ground school portion of such a course.

2. A written statement from an FAA certificated ground or flight instructor, certifying that the applicant has satisfactorily completed the ground instruction required for the certificate or rating sought.

3. Logbook entries certified by an FAA certificated ground or flight instructor, showing satisfactory completion of the ground instruction required for the certificate or rating sought.

4. A certificate of graduation or statement of accomplishment from a ground school course appropriate to the certificate or rating sought conducted by an agency such as a high school, college, adult education program, the Civil Air Patrol, or an ROTC Flight Training Program appropriate to the certificate or rating sought.

5. A certificate of graduation from an aviation home study course appropriate to the certificate or rating sought.

6. A written statement from an FAA certificated ground or flight instructor, certifying that he/she has personally reviewed the applicant's completion of an aviation home study course appropriate to the certificate or rating sought and has found that the person has satisfactorily completed this course.

7. An applicant who is unable to provide any of the above when applying for the test may present the aviation home study course he/she has completed. The inspector will review it and may question the applicant to determine that the course was completed.

NOTE:

Applicants are encouraged to obtain the necessary ground instruction as described in items 1 through 6. Those who elect to apply for a written test, as described in item 7, must have their qualifications reviewed at the Flight Standards District Office. This review will be conducted on an appointment basis only, due to the FAA inspector workload.

APPROVED SCHOOL GROUND TRAINING

FAR Part 141 lists not only the general subjects which must be included in most approved courses of training, but also specifies the number of hours of ground instruction an approved school applicant must receive for a pilot or flight instructor certificate or rating. The totals for the various courses are shown in the following list.

1.	Private Pilot Certification Course	35 hours
2.	Private Test Course	35 hours
3.	Instrument Rating Course	30 hours
4.	Commercial Pilot Certification Course	100 hours
5.	Commercial Test Course	50 hours
6.	Airline Transport Pilot Test Course	40 hours
7.	Flight Instructor Certification Course	40 hours
8.	Additional Flight Instructor Rating Course	20 hours

INSTRUCTOR ENDORSEMENTS

Flight instructors are allowed to provide endorsements for certain flying activities as well as recommend students for flight tests. Accompanying this privilege of endorsement is an obligation to the person being endorsed. Regardless of the endorsement or recommendation, it assures the applicant that at *least* all of the training requirements established in the FARs have been accomplished.

MEDICAL CERTIFICATE __THIRD__ AA-7094608
AND STUDENT PILOT CERTIFICATE __CLASS__

THIS CERTIFIES THAT *(Full name and address)*

BARNETT, STEPHEN SCOTT
23146 HUMBOLT AVE.
ANYTOWN, U.S.A. 000001

DATE OF BIRTH	HEIGHT	WEIGHT	HAIR	EYES	SEX
4/10/63	72"	160	BRN	BLU	M

has met the medical standards prescribed in Part 67, Federal Aviation Regulations for this class of Medical Certificate, and the standards prescribed in Part 61 for a Student Pilot Certificate.

STUDENT PILOTS ARE PROHIBITED FROM CARRYING PASSENGERS.

LIMITATIONS

HOLDER SHALL WEAR LENSES WHICH CORRECT FOR NEAR AND DISTANT VISION WHILE EXERCISING THE PRIVILEGES OF HIS AIRMAN CERTIFICATE.

DATE OF EXAMINATION	EXAMINER'S SERIAL NO.
8/18/__	12345-0

EXAMINER

SIGNATURE *Robert P. McDonald M.D.*

TYPED NAME ROBERT P. McDONALD

AIRMAN'S SIGNATURE *Stephen S. Barnett*

FAA FORM 8420-2

FRONT SIDE

CONDITIONS OF ISSUE: This certificate shall be in the personal possession of the airman, at all times while exercising the privileges of his airman certificate. As a medical certificate, it is temporary for a period of 60 days; as a student pilot certificate, it is temporary for a period of 90 days. If no notice to the contrary is received within such periods, it will remain in effect until the expiration dates as provided in Sections 61.9(a) and 61.43 of the Federal Aviation Regulations, unless modified or recalled by proper authority. The holder of this certificate is governed by the provisions of FAR Secs. 61.45, 63.19, and 65.45(c) relating to physical deficiency.

CERTIFICATED INSTRUCTOR'S ENDORSEMENTS FOR STUDENT PILOTS

I certify that the holder of this certificate has met the requirements of the regulations and is competent for the following:

BACK SIDE

	DATE	MAKE AND MODEL OF AIRCRAFT	INSTRUCTOR'S SIGNATURE	INSTRUCTOR'S CERTIFICATE NO.	EXP. DATE
A. TO SOLO THE FOLLOWING AIRCRAFT	1/21/__	C-152	*Bert Swenson*	1957527	4/30/__
B. TO MAKE SOLO CROSS-COUNTRY FLIGHTS	AIRCRAFT CATEGORY				
	AIRPLANE				
	GLIDER				
	ROTORCRAFT				

NOTICE: Any alteration of this certificate is punishable by a fine not exceeding $1,000, or imprisonment not exceeding 3 years, or both.

Fig. 3-2. Student Pilot Certificate and Solo Endorsement

TYPES OF ENDORSEMENTS

Instructor endorsements are required on student pilot certificates and pilot logbooks for certain operations. The following is a brief discussion of some key endorsements and their importance.

STUDENT PILOT CERTIFICATE ENDORSEMENTS

Before solo privileges may be exercised, each student pilot certificate must be endorsed for solo flights by the instructor who provided instruction for solo, and who administered the pre-solo written exam. In addition, the instructor who provided cross-country training also must make the initial cross-country endorsement on the student pilot certificate. Solo endorsements are probably the most important ones an instructor can make. Although a flight instructor may endorse a student certificate on the basis of the specific circumstances surrounding the first solo flight, or a proposed cross-country flight, this endorsement is, in effect, issuing a license for future solo operations by the student. Figure 3-2 illustrates a student pilot certificate endorsed for solo privileges.

Before making the endorsement, the instructor should carefully consider the student's qualifications and determine not only that the student meets the requirements of the Federal Aviation Regulations, but also that the student is competent to make future solo flights within the scope of the regulations. It also is a good policy for instructors to require notification from their students before they embark on any solo flight. In addition, FAR Part 141 requires that each training flight (dual or solo) include a preflight briefing and a postflight critique of the student by the instructor assigned to that flight.

LOGBOOK ENDORSEMENTS

FAR Part 61 lists the required endorsements for student pilot operations. Specific logbook endorsements include local solo, solo flights at airports within 25 nautical miles of the home base, solo cross-country privileges, and solo flights in Class B airspace. A separate endorsement is required for each Class B airspace area and each cross-country flight. Although a single endorsement may be issued for repeated specific solo cross-country flights that are within 50 nautical miles. Logbook endorsements normally should contain the date, type of endorsement and the type of aircraft. The individual solo cross-country endorsement should include the date of flight, route of flight, points of intended landing, and type of aircraft. Class B airspace endorsements must indicate that the required ground and flight instruction was given for that Class B airspace area by the endorsing instructor. Figure 3-3 illustrates logbook endorsements for solo within 25 nautical miles and for solo cross-country privileges. Some other logbook endorsements include the following:

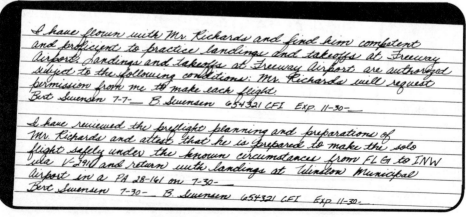

Fig. 3-3. Solo Logbook Endorsements

1. Any time dual instruction is given
2. When a person is recommended for a pilot certificate or additional rating
3. When a flight review or instrument competency check has been successfully completed
4. After successful completion of a flight check in a high performance aircraft
5. After a successful checkout in a tailwheel airplane
6. After successful completion of high altitude training

8710-1 APPLICATION

Each recommendation for a flight test must include a completed application for the certificate or rating as well as the appropriate logbook endorsement. In addition, it is *recommended* that the instructor sign the applicant's written test report stating the items missed on the written test have been reviewed; however, the instructor's signature on the appropriate line of the application for flight test will be accepted in lieu of this.

An example of a completed Airman Certificate and/or Rating Application is illustrated in figure 3-4. The reverse side of the form contains space for the instructor's signature and comments from the examining authority.

US Department of Transportation
Federal Aviation Administration

Airman Certificate and/or Rating Application

I Application Information ☐ Student ☐ Recreational ☒ Private ☐ Commercial ☐ Airline Transport ☐ Instrument
☐ Additional Aircraft Rating ☒ Airplane Single-Engine ☐ Airplane Multiengine ☐ Rotorcraft ☐ Glider ☐ Lighter-Than-Air
☐ Flight Instructor ___ Initial ___ Renewal ___ Reinstatement ☐ Additional Instructor Rating ☐ Ground Instructor
☐ Medical Flight Test ☐ Reexamination ☐ Reissuance of ______ Certificate ☐ Other ______

A. Name (Last, First, Middle): Anthony, Greg Michael
B. SSN (US Only): 524-68-1524
C. Date of Birth (Mo. Day Year): 08-20-75
D. Place of Birth: Denver, CO
E. Address (Please See Instructions Before Completing): 55 Inverness Drive East
City, State, Zip Code: Englewood, Colorado 80112
F. Nationality (Citizenship) Specify: ☒ USA ☐ Other ______
G. Do you read, speak and understand English? ☒ Yes ☐ No
H. Height: 70 in
I. Weight: 195 lbs
J. Hair: Brown
K. Eyes: Brown
L. Sex: ☒ Male ☐ Female
M. Do you now hold, or have you ever held an FAA Pilot Certificate? ☒ Yes ☐ No
N. Grade Pilot Certificate: Student
O. Certificate Number: BB7999090
P. Date Issued: 09-20-__
Q. Do you hold a Medical Certificate? ☒ Yes ☐ No
R. Class of Certificate: 3rd
S. Date Issued: 01-25__
T. Name of Examiner: Dr. Glenn C. Casch
U. Have you been convicted for violation of Federal or State statutes relating to narcotic drugs, marijuana, or depressant or stimulant drugs or substances ☐ Yes ☒ No
V. Date of Final Conviction
W. Glider or Free Balloon Pilots only: *Medical Statement: I have no known physical defect which makes me unable to pilot a glider or free balloon.* Signature X. Date

II Certificate or Rating Applied For on Basis of:

☐ A. Completion of Required Test — 1. Aircraft to be used (if flight test required): Cessna 172; 2a. Total time in this aircraft: 45.1 hours; 2b. Pilot in command: 17.5 hours

☐ B. Military Competence Obtained in — 1. Service; 2. Date Rated; 3. Rank or Grade and Service Number; 4. Has flown at least 10 hours as pilot in command during the past 12 months in the following military aircraft.

☒ C. Graduate of Approved Course — 1. Name and Location of Training Agency or Training Center: West Texas Wings of Flight, El Paso Int'l Airport, El Paso, Texas; 1a Certification Number: WE-00-199; 2. Curriculum From Which Graduated: Private Pilot Certificate Course; 3. Date: 06-29-__

☐ D. Holder of Foreign License Issued By — 1. Country; 2. Grade of License; 3. Number; 4. Ratings

☐ E. Completion of Air Carrier's Approved Training Program — 1. Name of Air Carrier; 2. Date; 3. Which Curriculum ☐ Initial ☐ Upgrade ☐ Transition

III Record of Pilot time *(Do not write in the shaded areas.)*

	Total	Instruction Received	Solo	Pilot in Command	Second in Command	Cross Country Instruction Received	Cross Country Solo	Cross Country Pilot in Command	Instrument	Night Instruction Received	Night Take-off Landing	Night Pilot in Command	Night Take-off Landing Pilot in Command	Number of Flights	Number of Aero-Tows	Number of Ground Launches	Number of Powered Launches	Number of Free Flights
Airplanes	45.1	27.6	17.5			4.5	10.3		3.7	3.1	12							
Rotorcraft																		
Gliders																		
Lighter than Air																		
Training Device Simulator																		

IV Have you failed a test for this certificate or rating? ☐ Yes ☒ No **Within the Past 30 days?** ☐ Yes ☒ No

V Applicant's Certification — I certify that all statements and answers provided by me on this application form are complete and true to the best of my knowledge, and I agree that they are to be considered as part of the basis for issuance of any FAA certificate to me. I have also read and understand the Privacy Act statement that accompanies this form.

Signature of Applicant: Greg M. Anthony Date: 9-24-__

FAA Use Only

EMP	REG	D.O.	SEAL	CON	ISS	ACT	LEV	TR	S.H.	SRCH	RTE		RATING (1)

FAA Form 8710-1 (7-92) Supersedes Previous Edition

Fig. 3-4. Airman Certificate and/or Rating Application

A flight instructor may recommend an individual to take an FAA written examination. This recommendation is proof that a course of ground instruction or a home study course has been completed. Under FAR Part 61, this proof is required prior to taking an examination. A flight instructor also may recommend an individual for re-testing after failure. This is required if the individual wishes to re-test within 30 days of failure. If an applicant fails a flight test, the instructor must sign a new application and, again, endorse the student's logbook. It is recommended the instructor state in the logbook endorsement that instruction has been given in the deficient areas, and that the applicant has been found to be competent and able to pass the flight test.

RECOMMENDED ENDORSEMENTS

The following examples are recommended for use by authorized instructors when endorsing logbooks, providing written statements for airmen applying for written or practical tests, and when certifying accomplishment of requirements for pilot operating privileges. Each endorsement should include the instructor's signature, date of signature, CFI certificate number, and certificate expiration date. A reference to the appropriate FAR is provided for each endorsement.

STUDENT PILOT ENDORSEMENTS

1. **Endorsement for presolo aeronautical knowledge:** FAR § 61.87(b)

Mr./Ms. _____ has satisfactorily completed a presolo written examination demonstrating knowledge of the portions of FAR Parts 61 and 91 applicable to student pilots, and the flight characteristics and operational limitations for a (make and model aircraft).
S/S (date) J.J.Jones 654321CFI Exp.___

2. **Endorsement for presolo flight training:** FAR § 61.87(c)

I have given Mr./Ms. _____ the fight instruction required by FAR § 61.87(c) in a (make and model aircraft). He/She has demonstrated proficiency in the applicable maneuvers and procedures listed in FAR § 61.87(d) through (j) [as appropriate] and is competent to make safe solo flights in a (make and model aircraft).
S/S (date) J.J.Jones 654321CFI Exp.___

3. **Endorsement for solo (each additional 90-day period):** FAR § 61.87(m)

I have given Mr./Ms. _____ the instruction required by FAR § 61.87(m). He/She has met the requirements of FAR § 61.87(m) and is competent to make safe solo flights in a (make and model aircraft).
S/S (date) J.J.Jones 654321CFI Exp.___

4. **Endorsement for solo landings and takeoffs at another airport within 25 n.m:** FAR 61.93(b)

I have flown with Mr./Ms. _____ and find him/her competent and proficient at practice landings and takeoffs at (airport name). Landings and takeoffs at (airport name) are authorized subject to the following conditions: (list applicable conditions).
S/S (date) J.J.Jones 654321CFI Exp.___

5. **Endorsement for each solo cross-country flight:** FAR § 61.93(d)(2)(i)

I have reviewed the preflight planning and preparations of Mr./Ms. _____ and attest that he/she is prepared to make the solo flight safely under the known circumstances from (location) to (destination) via (route of flight) with landings at (names of applicable airports) in a (make and model aircraft) on (date).
S/S (date) J.J.Jones 654321CFI Exp__

Note: The instructor may want to stipulate additional conditions in the above endorsement.

6. Endorsement for repeated solo cross-country flights not more than 50 nm from the point of departure: FAR § 61.93(d)(2)(ii)

I have given Mr./Ms. _____ flight instruction in both directions over the route between (airport name) and (airport name), including takeoffs and landings at the airports to be used, and find him/her competent to conduct repeated solo cross-country flights over that route, subject to the following conditions: (list applicable conditions).
S/S (date) J.J.Jones 654321CFI Exp.___

7. Endorsement for solo flight in Class B airspace: FAR § 61.95(a)

I have given Mr./Ms. _____ the ground and flight instruction required by FAR § 61.95(a)(1), and find him/her competent to conduct solo flight in (name of Class B airspace) Class B airspace.
S/S (date) J.J.Jones 654321CFI Exp.___

8. Endorsement for solo flight to, from, or at an airport located within Class B airspace: FAR §§ 61.95(b) and 91.131(b)(1)(ii)

I have given Mr./Ms. _____the ground and flight instruction required by FAR § 61.95(b)(1), and find him/her competent to conduct solo flight operations at (name of airport).
S/S (date) J.J.Jones 654321CFI Exp.___

RECREATIONAL PILOT ENDORSEMENTS

9. Endorsement for aeronautical knowledge: FAR §§ 61.35(a)(1) and 61.97

I certify that I have given Mr./Ms. _____ the ground instruction required by FAR § 61.97(a) through (h).
S/S (date) J.J.Jones 654321CFI Exp.___

10. Endorsement for flight proficiency: FAR § 61.98

I certify that I have given Mr./Ms. _____ the flight instruction required by FAR § 61.98(a)(1) through (8), and find him/her competent to perform each pilot operation safely as a recreational pilot.
S/S (date) J.J.Jones 654321CFI Exp.___

11. Endorsement for recreational pilot to act as PIC within 50 nm of the airport where instruction was received: FAR § 61.101(a)(3)(i)

I certify that I have given Mr./Ms. _____ the flight and ground instruction required by FAR § 61.101(a)(3)(i) at (name of airport).
S/S (date) J.J.Jones 654321CFI Exp.___

12. Endorsement for recreational pilot with fewer than 400 flight hours logged who has not logged PIC time within the preceding 180 days: FAR § 61.101(d)

I certify that I have given Mr./Ms. _____ flight instruction in a (make and model aircraft) and consider him/her competent to act as PIC of that aircraft.
S/S (date) J.J.Jones 654321CFI Exp.___

13. Endorsement for a recreational pilot to conduct solo flights for the purpose of obtaining an additional certificate or rating while under the supervision of an authorized flight instructor: FAR § 61.101(f)

I certify that I have given Mr./Ms. _____ the ground and flight instruction required by FAR § 61.87 in a (make and model aircraft). I find that he/she meets the aeronautical knowledge and flight training requirements of FAR § 61.87 and is competent to conduct a solo flight on (date) under the following conditions: (List all conditions which require en-

dorsement, e.g., flight which requires communication with ATC, flight between sunset and sunrise, flight in an aircraft for which the pilot does not hold a category/class rating).
S/S (date) J.J.Jones 654321CFI Exp.___

14. Endorsement for recreational pilot to act as PIC on a flight in excess of 50 nm for the purpose of obtaining an additional certificate or rating: FAR § 61.101(h)

I have given Mr./Ms. _____ the flight training required by FAR § 61.93 in a (make and model aircraft) and find him/her competent to conduct solo cross-country flights in that make and model aircraft, subject to the conditions stipulated in an endorsement for each specific cross-country flight.
S/S (date) J.J.Jones 654321CFI Exp.___

Note: The guidance in paragraph 5 may be used for an endorsement attesting to the review of the preflight planning and preparation of each specific cross-country solo flight.

PRIVATE PILOT ENDORSEMENTS

15. Endorsement for aeronautical knowledge: FAR §§ 61.35(a)(1) and 61.105(a)

I certify that I have given Mr./Ms. _____ the ground instruction required by FAR § 61.105(a)(1) through (6).
S/S (date) J.J.Jones 654321CFI Exp.___

16. Endorsement for flight proficiency: FAR § 61.107(a)

I certify that I have given Mr./Ms. _____ the flight instruction required by FAR § 61.107(a)(1) through (10) and find him/her competent to perform each pilot operation safely as a private pilot.
S/S (date) J.J.Jones 654321CFI Exp.___

COMMERCIAL PILOT ENDORSEMENTS

17. Endorsement for aeronautical knowledge: FAR §§ 61.35(a)(1) and 61.125(a)

I certify that I have given Mr./Ms. _____ the ground instruction required by FAR § 61.125(a)(1) through (4).
S/S (date) J.J.Jones 654321CFI Exp.___

18. Endorsement for flight proficiency: FAR § 61.127(a)

I certify that I have given Mr./Ms. _____ the flight instruction required by FAR § 61.127(a)(1) through (6) and find him/her competent to perform each pilot operation as a commercial pilot.
S/S (date) J.J.Jones 654321CFI Exp.___

INSTRUMENT RATING ENDORSEMENTS

19. Endorsement for aeronautical knowledge: FAR § 61.65(b)

I certify that I have given Ms./Ms. _____ the ground instruction required for the instrument (airplane/helicopter) rating by FAR § 61.65(b)(1) through (4).
S/S (date) J.J.Jones 654321CFI Exp.___

20. Endorsement for flight proficiency: FAR § 61.65(c) or (d)

I certify that I have given Mr./Ms. _____ the flight instruction in an (airplane/helicopter) required by FAR § 61.65(c)(d) and find him/her competent to perform each pilot operation as an instrument pilot.
S/S (date) J.J.Jones 654321CFI Exp.___

FLIGHT INSTRUCTOR ENDORSEMENTS

21. Endorsement for aeronautical knowledge: FAR § 61.185(a) and (b)

I certify that Mr./Ms. ____ has satisfactorily completed the course of instruction required by FAR § 61.185(a)(1) through (6), and has logged the ground instruction required by FAR §§ 61.105(a),

61.125(a) and 61.65(b) in preparation for the flight instructor (airplane/instrument) rating.
S/S (date) J.J.Jones 654321CFI Exp.___

22. Endorsement for Flight Proficiency: FAR § 61.187(a)

I certify that I have given Mr./Ms. _____ the flight instruction required by FAR § 61.187(a)(1) through (6) and find him/her competent to pass a practical test on those subjects.
S/S (date) J.J.Jones 654321CFI Exp.___

23. Endorsement for spin training: FAR § 61.187(a)(6)

I have given Mr./Ms. _____ flight training in spin entry, spins, and spin recovery techniques and he/she has demonstrated instructional competency in those maneuvers.
S/S (date) J.J.Jones 654321CFI Exp.___

Note: The above spin training endorsement is required of flight instructor-airplane and flight instructor-glider applicants only.

ADDITIONAL ENDORSEMENTS

24. Endorsement for completion of flight review: FAR § 61.56

Mr./Ms. _____, holder of pilot certificate #_____, has satisfactorily completed the flight review required by FAR § 61.56 on (date).
S/S (date) J.J.Jones 654321CFI Exp.___

Note: No logbook entry reflecting unsatisfactory performance on a flight review is required.

---

25. Endorsement for completion of an instrument competency check: FAR § 61.57(e)(2)

Mr./Ms. _____, holder of pilot certificate # _____, has satisfactorily completed an instrument competency check on (date).
S/S (date) J.J.Jones 654321CFI Exp.___

Note: No logbook entry reflecting unsatisfactory performance on an instrument competency check is required.

---

26. Endorsement for a pilot to act as PIC in a high performance airplane: FAR § 61.31(e)

I certify that I have given flight instruction in a high performance airplane to Mr./Ms. _____, holder of pilot certificate # _____, and find him/her competent to act as PIC in high performance airplanes.
S/S (date) J.J.Jones 654321CFI Exp.___

27. Endorsement for high altitude operations: FAR § 61.31(f)

I have given Mr./Ms. _____, holder of pilot certificate #_____, the ground and flight instruction on high altitude operations required by FAR § 61.31(f).
S/S (date) J.J.Jones 654321CFI Exp.___

28. Endorsement for a pilot to act as PIC in a tailwheel airplane: FAR § 61.31(g)

I have given Mr./Ms. _____, holder of pilot certificate #_____, flight instruction in normal and crosswind takeoffs and landings, wheel landings (if appropriate), and go-around procedures in a tailwheel airplane and find him/her competent to act as PIC in tailwheel airplanes.
S/S (date) J.J.Jones 654321CFI Exp.___

29. Endorsement for a pilot who does not hold an appropriate category/class rating to act as PIC of an aircraft in solo operations: FAR § 61.31(d)(2)

I certify that I have given Mr./Ms. _____ flight instruction in the pilot operations required for first solo in an (category and class of aircraft) and find him/her

competent to solo that category/class of aircraft.
S/S (date) J.J.Jones 654321CFI Exp.___

Note: The instructor may want to stipulate additional conditions (such as 90 day limitation) in the above endorsement.

---

30. **Endorsement to certify completion of prerequisites for a practical test:** FAR § 61.39(a)(5)

I have given Mr./Ms. ____ flight instruction in preparation for a (type of practical test) practical test within the preceding 60 days and find him/her competent to pass the test and to have satisfactory knowledge of the subject areas in which the applicant was shown to be deficient by his/her airman written test.
S/S (date) J.J.Jones 654321CFI Exp.___

Note: The instructor's signature in the endorsement block on the reverse side of FAA Form 8710-1, Airman Certificate and/or Rating Application, will be accepted in lieu of the above endorsement provided all appropriate FAR Part 61 requirements are substantiated by reliable records. However, the above endorsement without the instructor's signature in the endorsement block of FAA Form 8710-1 is not acceptable.

---

31. **Endorsement for retesting within 30 days of first failure of a written or practical test:** FAR § 61.49

I haven given Mr./Ms. ____ additional (flight/ground) instruction and find him/her competent to pass the (name of the written or practical test) test.
S/S (date) J.J.Jones 654321CFI Exp.___

Note: The instructor may also complete the endorsement in the space provided at the bottom of the applicant's airman written test report in

the case of a first failure on a written test. The instructor must sign the block provided for the instructor's endorsement on the reverse side of FAA Form 8710-1 for each retake of a practical test. An applicant may retake either a written or practical test within 30 days of a first failure if he or she has received additional instruction and an instructor's endorsement. For subsequent failures of both practical and written tests, the applicant must wait until 30 days after he/she failed the test before applying for retesting.

---

32. **Endorsement for an airman seeking an additional aircraft rating (other than airline transport):** FAR § 61.63(b) or (c)

I have given Mr./Ms. ____ the instruction required by FAR § 61.63 and find him/her competent to pass the (private/commercial) practical test for the addition of a (name of the applicable category/class) rating.
S/S (date) J.J.Jones 654321CFI Exp.___

FLIGHT INSTRUCTOR RECORDS

Flight instructors are required to sign the logbook of each person to whom they have given flight or ground instruction. The amount of time the instruction required and the date on which it was given also must be entered in the person's logbook. In addition, an instructor must keep a personal record of student pilot endorsements and test recommendations. The record usually is in the form of a pilot logbook, but it may be in a separate document. The record should contain the name of each person whose logbook or student pilot certificate has been endorsed for solo flight privileges and, in addition, the name of each person who has been recommended for a written, flight, or practical test, the type of test, result of the test, and the date of certification.

Instructors will be assigned additional record keeping duties when they are employed by an approved pilot school. FAR Part 141 requires that each holder of a pilot school or provisional pilot school certificate establish and maintain a current and accurate record of the participation and accomplishments of each student enrolled in an approved course of training conducted by the school (the student's logbook is not acceptable for this record). Additional requirements for record keeping in FAR Part 141 schools normally are fully explained to new instructors by the chief instructor of the school.

PRACTICING PROFESSIONALISM

The beginning flight instructor may have experienced a highly successful training program as evidenced by high scores on the FAA written examinations and flight tests. In addition, the new instructor may be a very affable, communicative individual with sincere interest and dedication in teaching people to fly. In fact, a few successful recommendations for flight tests may show that the instructor is performing above average. Unfortunately, without other efforts, these desirable accomplishments will not spell success. There are several other areas the instructor must concentrate on besides providing excellent instruction. They may be summed up in the phrase "practicing professionalism."

POTENTIAL STUDENTS

Every individual who makes an inquiry about learning to fly or obtaining an additional certificate or rating should be recognized as a *potential student*. Such a person may come to the airport personally to inspect the facility and equipment and determine the approximate cost of a planned training program. In other cases, the inquiry may be made by telephone or even in a personal letter. Regardless of the method used, such an inquiry provides the first opportunity for the instructor to practice professionalism. Time and again potential students

fail to develop because no one took the time to fully explain the ramifications of a training program and show a personal concern for their training ambitions. Some good rules for dealing with inquiries from potential students are shown in the following list.

1. Answer all questions fully and candidly.
2. Describe the full capabilities of the training facility and equipment.
3. Emphasize the beneficial aspects of training with the particular operator.
4. Introduce the potential student to other key people in the operation, if possible.
5. Be sure to obtain the potential student's address and telephone number for a follow-up contact.

Item 5 is most often neglected by beginning instructors, yet it is, by far, the most fundamental. Students seldom enroll immediately in a course of training after an initial inquiry. More than likely they have interviewed several operators and need some time to weigh general impressions of the inquiries they have made. What could be more timely during this decision period than to receive a phone call from an instructor who displays a personal interest in the training options available? Follow-up calls can be made within a few days to a few weeks, depending on the instructor's analysis of the individual's level of interest. They can be the deciding factor in a student's decision to begin a training program.

DROPOUTS

After training begins the instructor must not assume that the excitement of the training program alone will keep the student progressing toward the training goal. The instructor must continually strive to maintain student interest. Some of the actions that may be taken to achieve this are being prepared for each lesson, demanding an appropriate level of performance, previewing maneuvers in future lessons, and injecting some of the humor inherent in any flight training endeavor.

One of the sad facts of the general aviation industry's performance in past years is the number of people who begin flying annually but, for one reason or another, discontinue their training program. A certain number of dropouts are to be expected because a small minority of potential students are not suited to flying because of emotional or psychological reasons, and some do not initially realize the scope of commitments in terms of effort, time, and finances that flying requires. However, the number of dropouts could be altered significantly (and needs to be) by professional instructors who are willing to take a personal interest in their students.

TOOLS OF THE TRADE

Preparation is a fundamental part of the professional instructor's daily activity. In order to be prepared, the beginning instructor must accumulate many essential items which may be grouped together as "tools of the trade." For example, a set of current FARs, a reference library, appropriate FAA publications, and a complete set of student materials are necessary for effective instruction.

Some other important items are a telephone number listing of active or potential students and a notebook for the instructor's personal use containing test guides, training syllabi, and other items needed on a daily basis. A model of the training airplane can be very helpful for pre- and postflight briefings with student pilots. A personal copy of the pilot's operating handbook for the training aircraft and a flashlight also are beneficial items for instructors to acquire. A serviceable instrument hood, devices for covering certain flight instruments, a headset, and a subscription for instrument charts are desirable items for an instrument instructor. Instructors need not acquire all of these items before they begin instructing, but may obtain them gradually as experience is gained.

OPERATORS

Most beginning instructors will gain their initial experience from one of several hundred fixed-base operators throughout the country. Whether the particular operator conducts FAR Part 61 instruction exclusively or maintains a complete FAR Part 141 series of courses, instructors must understand that their performance directly affects the success and profitability of the school. Without the fixed-base operator, instructors could not possibly meet the demands of the industry. The FBO provides a valuable framework for instructors, including the facility itself, advanced training aids such as audiovisual programs, classrooms, a fleet of airplanes, maintenance personnel and facilities, telephones, scheduling, insurance, and so on. It is obvious that the independent instructor cannot approach the FBO's capability of providing essential services. Therefore, instructors who are employed by FBOs, should remain cognizant of the operator's investment in their behalf and make a dedicated effort to improve the effectiveness of the operation overall, rather than isolating themselves from management considerations.

SECTION B — REGULATIONS
(FAR PART 61, SUBPART G — FLIGHT INSTRUCTORS)

61.181   APPLICABILITY

This subpart prescribes the requirements for the issuance of flight instructor certificates and ratings, the conditions under which those certificates and ratings are necessary, and the limitations upon these certificates and ratings.

61.183   ELIGIBILITY REQUIREMENTS: GENERAL

To be eligible for a flight instructor certificate a person must —
(a)   Be at least 18 years of age;
(b)   Read, write, and converse fluently in English;
(c)   Hold —
    (1)   A commercial or airline transport pilot certificate with an aircraft rating appropriate to the flight instructor rating sought, and
    (2)   An instrument rating, if the person is applying for an airplane or an instrument instructor rating;
(d)   Pass a written test on the subjects in which ground instruction is required by § 61.185; and
(e)   Pass a practical test on all items in which instruction is required by § 61.187 and, in the case of an applicant for a flight instructor-airplane or flight instructor-glider rating, present a logbook endorsement from an appropriately certificated and rated flight instructor who has provided the applicant with spin entry, spin, and spin recovery training in an aircraft of the appropriate category that is certificated for spins, and has found that applicant competent and proficient in those training areas. Except in the case of a retest after a failure for the deficiencies stated in § 61.49(b), the person conducting the practical test may either accept the spin training logbook endorsement or require demonstration of the spin entry, spin, and spin recovery maneuver on the flight portion of the practical test.

Amend #90
eff 4-15-91

61.185   AERONAUTICAL KNOWLEDGE

(a)   Present evidence showing that he has satisfactorily completed a course of instruction in at least the following subjects:
    (1)   The learning process.
    (2)   Elements of effective teaching.
    (3)   Student evaluation, quizzing, and testing.
    (4)   Course development.
    (5)   Lesson planning.
    (6)   Classroom instructing techniques.
(b)   Have logged ground instruction from an authorized ground or flight instructor in all of the subjects in which ground instruction is required for a private and commercial pilot certificate, and for an instrument rating, if an airplane or instrument instructor rating is sought.

61.187   FLIGHT PROFICIENCY

(a)   An applicant for a flight instructor certificate must have received flight instruction, appropriate to the instructor rating sought in the subjects listed in this paragraph by a person authorized in paragraph (b) of this section. In addition, his logbook must contain an endorsement by the person who has given him the instruction certifying that he has found the applicant competent to pass a practical test on the following subjects:
    (1)   Preparation and conduct of lesson plans for students with varying backgrounds and levels of experience and ability.
    (2)   The evaluation of student flight performance.
    (3)   Effective preflight and postflight instruction.
    (4)   Flight instructor responsibilities and certifying procedures.
    (5)   Effective analysis and correction of common student pilot flight errors.
    (6)   Performance and analysis of standard flight training procedures and maneuvers appropriate to the flight instructor rating sought. For flight instructor-airplane and flight instructor-glider applicants, this shall include the satisfactory demonstration of stall awareness, spin entry, spins, and spin recovery techniques in an aircraft of the appropriate category that is certificated for spins.

Amend #90
eff 4-15-91

(b)   The flight instruction required by paragraph (a) of this section must be given by a person who has held a flight instructor certificate during the 24 months immediately preceding the date the instruction is given, who meets the general requirements for a flight instructor certificate prescribed in §61.183, and who has given at least 200 hours of flight instruction, or 80 hours in the case of glider instruction, as a certificated flight instructor.

61.189   FLIGHT INSTRUCTOR RECORDS

(a)   Each certificated flight instructor shall sign the logbook of each person to whom he has given flight or ground instruction and specify in that book the amount of time and the date on which it was given. In addition, he shall maintain a record in his flight instructor logbook, or in a separate document containing the following:

   (1)   The name of each person whose logbook or student pilot certificate he has endorsed for solo flight privileges. The record must include the type and date of each endorsement.

   (2)   The name of each person for whom he has signed a certification for a written, flight, or practical test, including the kind of test, date of his certification, and the result of the test.

(b)   The record required by this section shall be retained by the flight instructor separately or in his logbook for at least 3 years.

61.191   ADDITIONAL FLIGHT INSTRUCTOR RATINGS

The holder of a flight instructor certificate who applies for an additional rating on that certificate must —

(a)   Hold an effective pilot certificate with ratings appropriate to the flight instructor rating sought;

(b)   Have had at least 15 hours as pilot in command in the category and class of aircraft appropriate to the rating sought; and

(c)   Pass the written and practical test prescribed in this subpart for the issuance of a flight instructor certificate with the rating sought.

61.193   FLIGHT INSTRUCTOR AUTHORIZATIONS

(a)   The holder of a flight instructor certificate is authorized, within the limitations of that person's flight instructor certificate and ratings, to give the —

   (1)   Flight instruction required by this part for a pilot certificate or rating;

   (2)   Ground instruction or a home study course required by this part for a pilot certificate and rating;

   (3)   Ground and flight instruction required by this subpart for a flight instructor certificate and rating, if that person meets the requirements prescribed in § 61.187(b);

   (4)   Flight instruction required for an initial solo or cross-country flight;

   (5)   Flight review required in § 61.56 in a manner acceptable to the Administrator;

   (6)   Instrument competency check required in § 61.57(e)(2);

   (7)   Pilot-in-command flight instruction required under § 61.101(d); and

   (8)   Ground and flight instruction required by this part for the issuance of the endorsements specified in paragraph (b) of this section.

(b)   The holder of a flight instructor certificate is authorized within the limitations of that person's flight instructor certificate and rating, to endorse —

   (1)   In accordance with §§ 61.87(m) and 61.93(c) and (d), the pilot certificate of a student pilot the instructor has instructed authorizing the student to conduct solo or solo cross-country flights, or to act as pilot in command of an airship requiring more than one flight crew member;

   (2)   In accordance with §§ 61.87(m) and 61.93(b) and (d), the logbook of a student pilot the flight instructor has instructed, authorizing single or repeated solo flights;

   (3)   In accordance with § 61.93(d), the logbook of a student pilot whose preparation and preflight planning for a solo cross-country flight the flight instructor has reviewed and found adequate for a safe flight under the conditions the flight instructor has listed in the logbook;

   (4)   In accordance with § 61.95, the logbook of a student pilot the flight instructor has instructed authorizing solo flights in a Class B airspace area or at an airport within a Class B airspace area;

┌ Amend #92
└ eff 9-16-93

   (5)   The logbook of a pilot or another flight instructor who has been trained by the person described in paragraph (b) of this section, certifying that the pilot or other flight instructor is prepared for an operating privilege, a written test, or practical test required by this part;

   (6)   In accordance with §§ 61.57(e)(2) and 61.101(d), the logbook of a pilot the flight instructor has instructed authorizing the pilot to act as pilot in command;

   (7)   [Reserved]; and

   (8)   In accordance with §§ 61.101(g) and (h), the logbook of a recreational pilot the flight instructor has instructed authorizing solo flight.

61.195   FLIGHT INSTRUCTOR LIMITATIONS

The holder of a flight instructor certificate is subject to the following limitations:

(a)   *Hours of Instruction.* He may not conduct more than eight hours of flight instruction in any period of 24 consecutive hours.

(b)   *Ratings.* Flight instruction may not be conducted in any aircraft for which the flight instructor does not hold a category, class, and if appropriate, a type rating, on the flight instructor's pilot and flight instructor certificates.

(c) *Endorsement of Student Pilot Certificate.* He may not endorse a student pilot certificate for initial solo or solo cross-country flight privileges, unless he has given that student pilot flight instruction required by this Part for the endorsement, and considers that the student is prepared to conduct the flight safely with the aircraft involved.

(d) *Logbook Endorsement.* He may not endorse a student pilot's logbook—
  (1) For solo flight unless he has given that student flight instruction and found that student pilot prepared for solo flight in the type of aircraft involved;
  (2) For a cross-country flight, unless he has reviewed the student's flight preparation, planning, equipment, and proposed procedures and found them to be adequate for the flight proposed under existing circumstances; or
  (3) For solo flights in a Class B airspace area or at an airport within a Class B airspace area unless the flight instructor has given that student ground and flight instruction and has found that student prepared and competent to conduct the operations authorized.

Amend #92
eff 9-16-93

(e) *Solo Flights.* He may not authorize any student pilot to make a solo flight unless he possesses a valid student pilot certificate endorsed for solo in the make and model aircraft to be flown. In addition, he may not authorize any student pilot to make a solo cross-country flight unless he possesses a valid student pilot certificate endorsed for solo cross-country flight in the category of aircraft to be flown.

(f) *Instruction in Multiengine Airplane or Helicopter.* He may not give flight instruction required for the issuance of a certificate or a category, or class rating, in a multiengine airplane or a helicopter, unless he has at least 5 hours of experience as pilot in command in the make and model of that airplane or helicopter, as the case may be.

(g) *Recreational Pilot Endorsements.* The flight instructor may not endorse a recreational pilot's logbook unless the instructor has given that pilot the ground and flight instruction required under this Part for the endorsement and found that pilot competent to pilot the aircraft safely.

61.197 RENEWAL OF FLIGHT INSTRUCTOR CERTIFICATES

The holder of a flight instructor certificate may have his certificate renewed for an additional period of 24 months if he passes the practical test for a flight instructor certificate and the rating involved, or those portions of that test that the Administrator considers necessary to determine his competency as a flight instructor. His certificate may be renewed without taking the practical test if —

(a) His record of instruction shows that he is a competent flight instructor;

(b) He has a satisfactory record as a company check pilot, chief flight instructor, pilot-in-command of an aircraft operated under Part 121 of this chapter, or other activity involving the regular evaluation of pilots, and passes any oral test that may be necessary to determine that instructor's knowledge of current pilot training and certification requirements and standards; or

Amend #95
eff 4-13-94

(c) He or she has successfully completed, within 90 days before the application for the renewal of his or her certificate, an approved flight instructor refresher course consisting of ground or flight instruction, or both.

61.199 EXPIRED FLIGHT INSTRUCTOR CERTIFICATES AND RATINGS

(a) *Flight Instructor Certificates.* The holder of an expired flight instructor certificate may exchange that certificate for a new certificate by passing the practical test prescribed in §61.187.

(b) *Flight Instructor Ratings.* A flight instructor rating or a limited flight instructor rating on a pilot certificate is no longer valid and may not be exchanged for a similar rating or a flight instructor certificate. The holder of either of those ratings is issued a flight instructor certificate only if he passes the written and practical test prescribed in this subpart for the issue of that certificate.

61.201 CONVERSION TO NEW SYSTEM OF FLIGHT INSTRUCTOR RATINGS

General. The holder of a flight instructor certificate that does not bear any of the new class or instrument ratings listed in §61.5(c)(2), (3), or (4) for a flight instructor certificate, may not exercise the privileges of that certificate. The holder of a flight instructor certificate with a glider rating need not convert that rating to a new class rating to exercise the privileges of that certificate and rating.

chapter 4

THE INSTRUMENT FLIGHT INSTRUCTOR

INTRODUCTION

Instrument instruction has its own techniques and procedures which require specialized knowledge and proficiency on the part of the instructor. This chapter is designed for those individuals who wish to add instrument instructing privileges to their flight instructor certificates. Consequently, it is assumed that the applicant is at least a commercial pilot with an instrument rating who also possesses a basic flight instructor certificate with airplane, single-engine land privileges. Some of the material presented is considered to be a review of instrument knowledge and procedures while the rest is oriented to the needs of the beginning instrument instructor. The first section includes a detailed analysis of the flight instruments followed by a thorough discussion of attitude instrument flying and basic navigation. The second section covers the full range of instrument procedures including IFR takeoffs, departures, holding patterns, approaches, and cross country from the viewpoint of the instrument instructor.

SECTION A—BASIC INSTRUMENT INSTRUCTION

The foundation for successfully teaching attitude instrument flying involves more than a knowledge of controlling the airplane by instrument reference. The theory of instrument flight is based partly on the use, function, and limitations of individual instruments. This is why a working knowledge of flight instruments is essential for both instrument instructors and students.

arranged logically to promote rapid scanning and built-in system redundancy prevents the failure of one instrument or component from causing complete loss of attitude reference. To further promote safety, a fundamental part of instrument training involves teaching the student the capabilities and limitations of flight instruments and supporting systems.

FLIGHT INSTRUMENTS

The instrumentation of modern light aircraft provides many safety features for the instrument pilot. Panel layouts are

GYROSCOPIC INSTRUMENTS

The *gyroscopic instruments* include the *heading indicator*, the *attitude indicator*, and the *turn coordinator*. Each of these instruments contains a gyro rotor driven by air or electricity and each makes use of gyroscopic principles to display the attitude of the aircraft. It is important for instrument instructors and students to understand gyroscopic instruments and the principles governing their operation.

The primary trait of a spinning gyro rotor is *rigidity in space*, or *gyroscopic inertia*. Newton's First Law states in part, "A body in motion tends to move in a constant direction and speed unless disturbed by some external force." The spinning rotor inside a gyroscopic instrument maintains a constant attitude in space as long as no outside forces are applied. This quality of stability is enhanced if the rotor has great mass and speed.

Gyroscopic inertia depends on the following design factors:
1. **Weight** — For a given size, a heavier mass is more resistant to disturbing forces than a lighter mass.
2. **Rotation Speed** — The higher the rotational speed, the greater the rigidity.
3. **Radius at Which the Weight is Concentrated** — Maximum rigidity is obtained from a mass when its principal weight is concentrated near the outside rim of the rotor and is rotated at high speeds.
4. **Bearing Friction** — Any friction applies a deflecting force to a gyro.

Two types of mountings are used. The free-mounted gyro is set on three gimbals with the gyro free to rotate in any plane, as shown in figure 4-1. The semirigid mounting utilizes two gimbals and the gyro is limited to two planes of rotation. The heading and attitude indicators are of the free-mounted type. The turn indicator employs the semirigid mounting.

Another characteristic of gyros is *precession*, which is the tilting or turning of the gyro axis as a result of applied forces. When force is applied to the rim of a

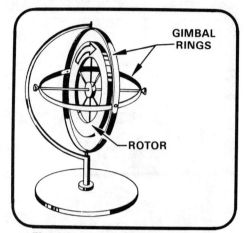

Fig. 4-1. Gyro with Free Mounting

gyro rotor that is not turning, the rotor naturally moves in the direction of the force. However, when the rotor is spinning, the same applied force will cause the rotor to move as though the force had been applied *at a point 90° around the rim in the direction of rotation*, as shown in figure 4-2. Unavoidable precession is caused by maneuvering and internal friction of attitude and directional gyros.

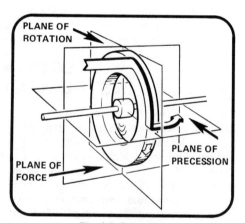

Fig. 4-2. Precession

POWER SOURCES

The gyroscopic instruments may be operated by either electrical power, a vacuum system, or both. In most light aircraft, the heading and attitude indicators are driven by the vacuum system and the turn coordinator is powered by the electrical system. This arrangement precludes the loss of all three instruments should one power source fail.

Aircraft that normally operate at high altitudes do not utilize a vacuum system to power flight instruments since vacuum pump efficiency is limited in the thin air at high altitudes. Many of these high altitude systems depend entirely on the aircraft electrical system to supply the power required for the gyroscopic instruments.

Vacuum System

Air-driven gyros normally are powered by a vacuum pump attached to and driven by the engine. Suction lines connect the pump to the instruments and *cabin air* is drawn through an air filter and then through the instrument case. As air enters the case, it is accelerated and directed against small "buckets" cast into the gyro wheel. A regulator is located between the pump and the gyroscopic in-

struments to regulate suction pressure as shown in figure 4-3. Since a constant gyro speed is essential for reliable instrument indications, correct suction pressure must be maintained.

Rotor speeds range from 8,000 to 18,000 r.p.m., depending on the make and type of instrument. Because of these operational speeds, time must be allowed for rotor acceleration to full speed before accurate indications can be expected. Therefore, a student should not be allowed to set the heading indicator immediately after engine start.

The vacuum system should provide a minimum suction of approximately four inches of mercury and a maximum of five inches; however, this will vary somewhat with different aircraft and instruments. The pilot's operating handbook should be consulted for the specific information for a particular model. Low gyro speeds result in slow instrument response or lagging indications, while fast gyro speeds cause the instruments

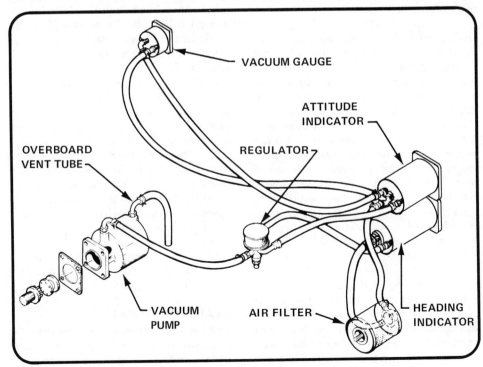

Fig. 4-3. Vacuum System

to overreact in addition to wearing the gyro bearings faster and decreasing gyro life.

THE ATTITUDE INDICATOR

The attitude indicator is the only instrument that provides an immediate and direct presentation of the aircraft pitch and bank attitude. It must be emphasized to the student that it depicts attitude and not necessarily performance. For example, it may show a nose-high attitude when the aircraft is not necessarily climbing. Only by cross-checking the other flight instruments can the actual performance be confirmed.

OPERATIONAL ERRORS

There are certain errors inherent in all gyroscopic instruments. The attitude indicator is subject to precession errors, usually caused by worn or dirty gimbal bearings. Low vacuum or a loss of vacuum also will prevent proper operation. Perhaps the most important error is that of acceleration/deceleration, particularly in high performance airplanes.

During rapid acceleration on takeoff, for example, the attitude indicator will show a higher pitch attitude than actually exists. This occurs because the horizon bar tends to move down during acceleration. A pilot correcting for such an erroneous indication would relax back pressure and a lower climb rate or possibly a descent would result. Therefore, takeoffs under low visibility conditions require the use of other instruments to confirm that a positive rate of climb has been established immediately after takeoff. Deceleration error has the opposite effect, causing the horizon bar to move up indicating a false, nose-low attitude.

Centrifugal force in a turn can cause the attitude indicator to precess, creating errors in both pitch and bank. These errors are usually minor and result in deviations of no more than five degrees of bank and one bar-width of pitch. The effect is greatest in a 180° steep turn. For example, after roll out from a 180° steep turn to straight-and-level flight by visual reference, the attitude indicator will show a slight climb and turn in the opposite direction. At the end of a 360° turn, the precession induced during the first 180° is cancelled out by precession in the opposite direction during the second 180° of turn. A skidding turn precesses the gyro toward the inside of the turn. After the aircraft returns to straight-and-level, coordinated flight, the miniature aircraft shows a turn in the direction opposite the skid.

OPERATIONAL LIMITATIONS

Many currently manufactured attitude indicators have different limitations than the older types. These instruments do not "tumble" when gyro limits are exceeded. They are designed to reflect bank throughout 360° of roll. However, they do not reflect pitch attitudes beyond 85° nose up or nose down. Beyond this limit, the newer gyros will give incorrect indications. A self-erecting mechanism is used to return the gyro to its correct position once a normal attitude is resumed. Older instruments have bank limits of 100° to 110° and pitch limits of 60° to 70°. If the aircraft attitude exceeds either limit, the indicator will topple over, rather than merely precess. This "tumbling" or "spilling" should be avoided since it will damage bearings and render the instrument useless until the gyro is erected again. For example, the rotor will re-erect at the rate of approximately eight degrees per minute. If the gyro tumbled at 80°, it could take as long as 10 minutes for the instrument to right itself. In order to prevent tumbling or to reset the gyro, some older gyros have caging devices. When "caged," the instrument cannot be used for attitude reference. Instructors should be familiar with the type of gyro instruments used in the training airplane and operate within the limitations specified.

HEADING INDICATOR

The heading indicator uses the principle of gyroscopic rigidity to provide a stable heading reference. In newer heading indicators, the vertical card or dial on the instrument face appears to revolve as the airplane turns. The heading is displayed at the top of the dial at the nose of the miniature airplane. Another type of direction indicator shows the heading on a ring which is similar to the card in a magnetic compass.

Modern heading indicators will tumble when pitch or bank exceeds 80°, while the older vintage instruments have limits of 55° pitch or bank. Like the attitude indicator, the rotational speed of the heading indicator is very high, usually 10,000 r.p.m. In addition, the vacuum pressure must be within the limits specified by the manufacturer for proper operation.

OPERATIONAL ERRORS

Since the heading indicator does not have direction-seeking capability, it should be checked and set by reference to the magnetic compass prior to takeoff. While airborne, precession causes the heading indicator to gradually drift away from the correct magnetic heading. Therefore, it should be checked and reset against the magnetic compass at least every 15 minutes. An error of three degrees in 15 minutes is acceptable for normal operations.

Two pilot-induced errors of the heading indicator are caused by incorrect procedures for setting the instrument. They are: not applying magnetic deviation to the magnetic compass reading when resetting the heading indicator; and attempting to obtain an accurate magnetic compass indication in any attitude other than straight-and-level, unaccelerated flight. The instructor must ensure that each student uses the correct procedures for setting the instrument.

TURN INDICATORS

One of the main functions of the turn coordinator and the turn-and-slip indicator is to permit the pilot to establish and maintain standard-rate turns. A standard-rate turn is a turn at a rate of three degrees per second. At this rate, a complete 360° turn takes two minutes. The bank required is directly related to true airspeed (TAS).

As TAS decreases, the angle of bank must decrease to maintain a standard-rate turn. Conversely, an increase in true airspeed requires an increase in bank to maintain standard rate. The appropriate bank angle for a standard-rate turn can be determined by dividing the true airspeed in knots by a factor of 10 and adding 5° to the result. At a true airspeed of 110 knots (11 + 5), a 16° angle of bank is required.

As shown in figure 4-4, the gyros in both the turn coordinator and turn-and-slip indicator are mounted so they rotate in the vertical plane. The gimbal in the turn coordinator is set at an angle, or canted, which means precession allows the gyro to sense both rate of roll and rate of turn. The gimbal in the turn-and-slip indicator is horizontal. In this case, precession allows the gyro to sense only rate of turn. When the miniature airplane or needle is aligned with the turn index, the aircraft is

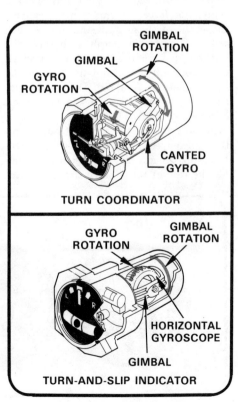

TURN COORDINATOR

TURN-AND-SLIP INDICATOR

Fig. 4-4. Turn Indicators

in a standard-rate turn. Since a coordinated turn requires a bank, the turn coordinator provides an indirect indication of bank.

Another part of the turn coordinator is the inclinometer. The position of the ball defines the quality of the turn, or whether the angle of bank is correct for the rate of turn. As shown in figure 4-5, the ball is centered in a coordinated turn (instrument 1). In a skid (instrument 2), the rate of turn is too great for the angle of bank, and the ball moves to the outside of the turn. Conversely, in a slip (instrument 3), the rate of turn is too small for the angle of bank, and the ball moves to the inside of the turn. In both cases, either the angle of bank or the rate of turn must be adjusted.

The inclinometer shows the relationship between opposing horizontal forces in a turn. The horizontal component of lift causes an airplane to turn and opposes centrifugal force. The inclinometer reflects the state of these opposing forces. For example, during a coordinated turn, the horizontal component of lift is balanced by centrifugal force, and the ball in the inclinometer is centered. During a skid, centrifugal force exceeds the horizontal component of lift. This makes the rate of turn too great for the angle of bank, and the ball moves to the outside of the turn. In a slip, the aircraft is banked too much for the rate of turn. Since the horizontal

component of lift exceeds centrifugal force, the ball moves to the inside of the turn.

Slipping or skidding also alters the normal load factor experienced in turns. This happens because the wings must generate enough lift to support the weight of the airplane and overcome centrifugal force. Since a skid generates a higher-than-normal centrifugal force, load factor is increased. In a slip, load factor decreases because centrifugal force is lower than normal. When the indicator is stabilized on the standard-rate turn index, the airplane is turning at three degrees per second, regardless of the position of the ball.

PITOT-STATIC INSTRUMENTS

The pitot-static instruments are the airspeed indicator, altimeter, and vertical speed indicator. All of these instruments operate on the principle of differential air pressure. Pressure sensitive mechanisms convert pressure supplied by the pitot-static system to a measurement of aircraft speed, altitude, and vertical speed. Pitot pressure, also called ram or dynamic pressure, is directed only to the airspeed indicator, but static (ambient) pressure is directed to all three instruments. Figure 4-6 provides a schematic diagram of a basic pitot-static system with electric pitot heat.

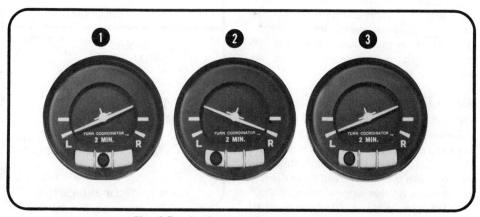

Fig. 4-5. Angle of Bank vs Rate of Turn

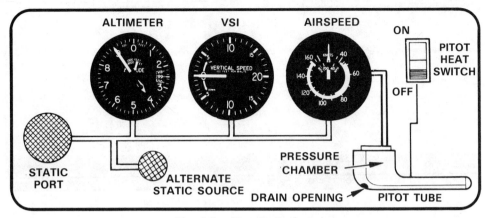

Fig. 4-6.   Pitot-Static System

AIRSPEED INDICATOR

Airspeed is measured by comparing the difference between the pitot and static pressures and, through mechanical linkages, displaying the resultant on the airspeed indicator. Three types of airspeed need to be considered by the instrument student.

Indicated airspeed (IAS) is read directly from the dial of the airspeed indicator. It equals true airspeed only at sea level on a standard day.

Calibrated airspeed (CAS) is indicated airspeed corrected for errors introduced by the pitot-static system. An airspeed correction table is provided by the aircraft manufacturer when the difference between indicated airspeed and calibrated airspeed is significant.

True airspeed (TAS) is the actual speed of the aircraft through the air. It is found by correcting calibrated airspeed for nonstandard temperature and pressure. The airspeed indicator is calibrated at standard sea level conditions. As an aircraft climbs and temperature and pressure decrease, the airspeed indicator reads slower than the aircraft actually is flying. Therefore, indicated airspeed normally is slower than true airspeed.

V-SPEEDS AND AIRSPEED COLOR MARKINGS

Colored arcs and radials on the face of the airspeed indicator define aircraft speed limits and operating ranges, as shown in figure 4-7. The following is a list of commonly used V-speeds and color codings.

V_{NE} — Never exceed (red line)

V_{NO} — Maximum for normal operations (high-speed end of green arc)

V_{FE} — Maximum with flaps extended (high-speed end of white arc)

V_{S1} — Stall, power-off, clean configuration (low-speed end of green arc)

V_{SO} — Stall, power-off, landing configuration (low-speed end of white arc)

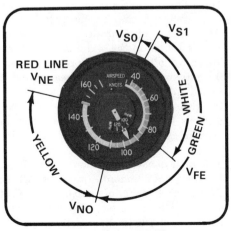

Fig. 4-7.   Airspeed Indicator

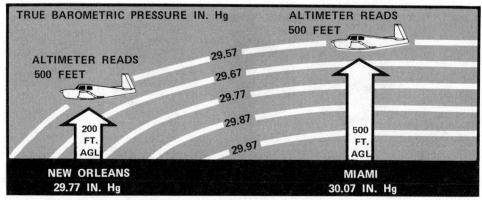

Fig. 4-8.   Pressure Effects on Altimeter

During gusty or turbulent atmospheric conditions, the aircraft should be flown at or below the design maneuvering speed (V_A) but with sufficient margin to avoid a stall. This important limiting speed normally is not marked on the airspeed indicator, but it is listed in the pilot's operating handbook and/or shown on a placard on the instrument panel near the airspeed indicator.

Two other important limiting speeds for airplanes with retractable landing gear are the maximum landing gear extended speed (V_{LE}) and the maximum landing gear operating speed (V_{LO}). Both of these are found in the pilot's operating handbook.

ALTIMETER

The altimeter senses the normal decrease in air pressure that accompanies an increase in altitude. The airtight instrument case is vented to the static port. With an increase in altitude, the air pressure within the interior of the case decreases and a sealed aneroid barometer (bellows) within the case expands. The barometer movement is transferred to the indicator via mechanical linkage.

The instrument calibration is based on International Standard Atmospheric (ISA); that is, 29.92 inches of mercury at 15°C at sea level. Since these conditions rarely exist, the altimeter must be adjusted to current local pressure settings if it is to provide accurate information. For this reason, it is imperative that the pilot adjust the altimeter to the most current setting along the route of flight.

Since nonstandard temperatures also affect altimeter accuracy, flight at a constant indicated level does not necessarily mean the aircraft is actually at a constant altitude above sea level. It should be recalled that flight from a warm airmass to a cold airmass will cause the altimeter to indicate an altitude higher than the actual flight level. Likewise, flight from an area of high pressure to one of low pressure will produce the same altimeter error. Figure 4-8 depicts pressure effects on altimeter indications. When terrain clearance is a factor in route selection, these inherent errors must be considered. The effects of nonstandard conditions can result in a difference between indicated and true altitude of 1,000 feet or more.

There are five definitions of altitude the instrument instructor must understand and explain to the student.

1. *Indicated altitude* is read directly from the altimeter when it is set to the current altimeter setting.

2. *Pressure altitude* is read from the altimeter when it is set to the standard atmosphere sea level pressure of 29.92 inches of mercury.

3. *Density altitude* is pressure altitude corrected for nonstandard temperature.

4. *True altitude* is the actual height above mean sea level.
5. *Absolute altitude* is the actual height above the earth's surface.

VERTICAL SPEED INDICATOR

The vertical speed indicator (VSI) measures how fast the ambient air pressure increases or decreases as the aircraft climbs or descends. The rate of change of static pressure is obtained by furnishing static pressure directly to a thin metallic diaphragm and through a calibrated orifice to an airtight case surrounding the diaphragm. As the aircraft climbs, the static pressure in the case is momentarily trapped by the calibrated orifice and the static pressure in the diaphragm is allowed to decrease immediately. This causes the diaphragm to contract. Through a mechanical linkage, the pointer indicates a climb. In a descent, the opposite is true. Case static pressure is momentarily trapped at the lower static pressure while the diaphragm expands because of the higher static pressure furnished. This expansion causes the pointer to indicate a descent.

As the static pressure in the diaphragm changes during entry to a climb or descent, the instrument will show an immediate change of vertical direction. In other words, it shows a trend. However, it requires six to nine seconds for the pressures in the instrument to stabilize and provide reliable rate information.

PITOT-STATIC SYSTEM BLOCKAGE

The normally high reliability of the pitot-static instruments can be seriously affected by blockages in the system. Blockages may be caused by moisture (including ice), dirt, or even insects. Careful visual inspection of pitot tube and static port openings may reveal blockages that require removal by a certificated mechanic. It also is possible for the pitot tube to become blocked when the aircraft is operating in visible moisture near the freezing level. This is why pitot heat should always be turned on during flight in visible moisture.

PITOT BLOCKAGE

The airspeed indicator is the only instrument affected by the pitot tube blockage. The system can become clogged in one of two ways. First, the ram air inlet can clog while the drain hole remains open. In this situation, the pressure in the line to the airspeed indicator will vent out the drain hole, causing the airspeed indicator to drop to zero. Often, this occurs when ice forms over the ram air inlet. The second situation occurs when both the ram air inlet and drain hole become clogged. When this occurs, the air pressure in the line is trapped. The airspeed indicator will no longer indicate changes in airspeed. If the static port is open, the indicator will react the same as an altimeter, showing an increase in airspeed as altitude increases and a decrease in speed as altitude decreases.

STATIC BLOCKAGE

If the static system is clogged, the airspeed indicator will continue to react to changes in airspeed, since ram air pressure is still being supplied by the pitot tube. However, the readings will not be correct. When operating above the altitude where the static port became clogged, the airspeed will read lower than it should. Conversely, when operating below that altitude, the indicator will read higher than the correct value. Since the altimeter determines altitude by measuring ambient air pressure, any blockage of the static port will "freeze" the altimeter in place and make it unusable. The VSI freezes at zero, since its only source of pressure is from the static port.

In many aircraft, an alternate static source is provided as a backup for the main static source. In nonpressurized aircraft, the alternate source usually is located inside the aircraft cabin. Due to the slipstream, the pressure inside the cabin is less than that of outside air for many aircraft. With the alternate static source selected, the altimeter will read a little higher and the airspeed a little faster than when using the normal static source. Also, the vertical speed will show a momentary climb as the system changes to the alternate source. In other aircraft or in specific aircraft configurations, the cabin pressure is slightly higher than outside air pressure so the altimeter reads lower with the alternate static source selected. The airspeed may also read lower than it should, and the vertical speed indicator may indicate a momentary descent after selecting the alternate source. If the aircraft is not equipped with an alternate static source, the glass of the vertical speed indicator can be broken to allow ambient air pressure to enter the static system. Of course, this makes the VSI unreliable for instrument reference. Do not use this technique on pressurized aircraft until you depressurize the cabin.

MAGNETIC COMPASS

The magnetic compass is a very simple self-contained instrument which requires no external power. It consists of a sealed outer case within which is located a pivot assembly and a float containing two or more magnets. The construction of the magnetic compass is shown in figure 4-9. The case is filled with an acid-free white kerosene which helps to dampen oscillations of the float and lubricate the pivot assembly. In addition, the pivot assembly

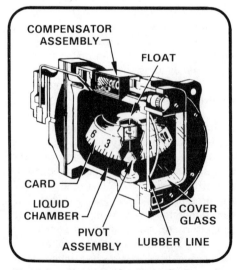

COMPENSATOR ASSEMBLY
FLOAT
CARD
LIQUID CHAMBER
COVER GLASS
PIVOT ASSEMBLY
LUBBER LINE

Fig. 4-9.   Magnetic Compass Construction

is spring-mounted to further dampen aircraft vibrations so the compass heading may be read more easily. A glass face is mounted on one side of the compass case with a lubber or reference line in the center. Compensating magnets are located within the case for use in calibrating the instrument. Most compasses have a small light powered from the electrical system to provide illumination of the instrument for night operations. The magnetic compass can display reliable directional guidance if its limitations and inherent errors (deviation, variation, magnetic dip, and oscillation) are understood.

DEVIATION AND VARIATION

Deviation is the deflection of the compass needle from a position of magnetic north as a result of local magnetic disturbances in the aircraft. The amount of deviation also varies with the operation of the aircraft's electrical equipment. Since these disturbances change with the passage of time, the compass must be swung (calibrated) periodically to compensate for these errors. Magnetic compasses present directional information in terms of magnetic north. To convert directions expressed in terms of true north to magnetic north, the angular difference between true and magnetic north (variation) is applied. Easterly variation is subtracted from and westerly variation is added to the true direction to determine the magnetic direction.

MAGNETIC DIP

Magnetic dip is the tendency of the compass needle to point downward as it moves closer to one of the magnetic poles. This tendency causes errors in compass readings during turns, acceleration, and deceleration. Acceleration and deceleration errors occur during airspeed changes and are most apparent on headings of east and west. When the aircraft is accelerating on either of these headings, a turn to the north is indicated. An aid for remembering this relationship between acceleration and deceleration is the acronym ANDS, or "accelerate north, decelerate south."

Northerly turning error is most apparent when turning to or from a heading of north or south. When making a turn from a northerly heading in the Northern Hemisphere, the compass gives a brief indication of a turn in the opposite direction. Then it *lags* behind the actual heading until roll-out or until the heading approaches east or west. Very little northerly turning error is apparent on an easterly or westerly heading. When making a turn from a southerly heading, the compass gives an indication of a turn in the correct direction, but at a much faster rate. The magnetic compass heading then *leads* the actual aircraft heading. Oscillation errors are caused by rough air or rough pilot techniques which cause the compass to swing back and forth, making it difficult to interpret accurately. The fluid in the compass dampens these oscillations to some extent so that in smooth air during straight-and-level, unaccelerated flight, the compass will indicate accurately.

ATTITUDE INSTRUMENT FLYING

Attitude instrument flying is a fundamental method for controlling an airplane by reference to instruments. It is based on an understanding of the flight instruments and systems and the development of the skills required to interpret and translate the information presented by the instruments into precise airplane control. The primary/support method of attitude instrument flying regards the attitude of the airplane as a function of pitch, bank, and power control. For a given maneuver, there are specific pitch, bank, and power instruments that should be used to control the airplane and obtain the desired performance. Those instruments which provide the most pertinent and essential information during a given condition of flight are termed *primary* instruments. Those which back up and supplement the primary instruments are termed *supporting* instruments. This is the reason the method is termed the *primary/support concept* of attitude instrument flying. It is

recommended by the FAA for instrument training conducted in light aircraft with low operating speeds.

There is another method for teaching attitude instrument flying which is based on the control and performance concept. It assumes that aircraft performance depends on how the attitude and power relationships of the airplane are controlled. It designates certain flight instruments as control and others as performance instruments. In general, this method relies heavily on the attitude indicator during most maneuvers and is particularly well suited for high performance turbojet aircraft.

The primary/support method of attitude instrument flying on which the following discussion is based does not lessen the value of any individual instrument. The attitude indicator, for example, is still the instrument which provides basic attitude reference. Since it is the only instrument that provides instant and direct aircraft attitude information, it should be considered primary during any change in pitch or bank attitude. After the new attitude is established, other instruments become primary, and the attitude indicator usually becomes a supporting instrument.

Instrument scan, instrument interpretation, and airplane control are the three fundamental skills involved in all instrument flight maneuvers.

INSTRUMENT SCAN

Instrument scan is probably the most important single factor in precise instrument flight. A good scan involves systematically referring to the proper instrument at the proper time to obtain the necessary information for the maneuver in progress. During a climbing turn, for example, the attitude indicator, turn coordinator, and airspeed indicator should receive the major emphasis. In straight-and-level flight, the airspeed indicator and the turn coordinator need to be checked only periodically to verify the indications of the other flight instruments.

The instrument student must learn to interpret each instrument at a glance, correlate each bit of information, and then apply appropriate control forces to maintain the desired attitude. Generally, the major problems in developing scan technique are *fixation, emphasis,* and *omission.*

FIXATION

Fixation results from a natural human inclination to observe a specific instrument accurately. Fixation on a single instrument usually results in poor control. For example, while performing a medium bank or shallow bank turn, the student may have a tendency to watch the turn coordinator throughout the turn, instead of including other instruments in the crosscheck. This fixation on the turn coordinator often leads to a loss of altitude through poor pitch and bank control.

EMPHASIS

Instead of relying on a combination of instruments necessary for airplane performance information, the instrument student sometimes places too much emphasis on a single instrument. This differs from fixation in that the student is using other instruments for information, but is devoting too much attention to a particular instrument.

OMISSION

During performance of a maneuver, the student sometimes fails to anticipate significant instrument indications following marked attitude changes. For example, during leveloff from a climb or descent, emphasis may be placed on pitch control instruments, while omitting the instruments which supply heading or roll information. Such omissions result in erratic control of heading and bank.

INSTRUMENT INTERPRETATION

The rapid and correct interpretation of each instrument is as important as the scan. The instrument instructor must be constantly alert for this problem. Often,

misinterpretation of the instruments goes hand in hand with one or more of the problems of fixation, emphasis, or omission. To overcome this problem the student must be taught why each instrument is used for a particular maneuver and what information is required.

AIRCRAFT CONTROL

The instrument flight instructor must emphasize the importance of smooth, coordinated control pressures. Rough, jerky, and heavy control movements lead to erratic and unstabilized flight. The instructor must also insist on proper trim techniques as an aid in smooth, precise control of the aircraft. If these techniques are not emphasized, the student will tire quickly and become even more erratic. This will lead to slow, discouraging progress.

The instructor must be constantly alert for signs of fatigue in the student. As a general rule, the beginning instrument student's flight lesson should not exceed one hour. As the student gains confidence and proficiency, the duration of the lessons may be extended.

STRAIGHT-AND-LEVEL FLIGHT

Straight-and-level flight is the first maneuver introduced to the instrument student. Reasonable proficiency in straight-and-level flight should be gained before other maneuvers are introduced. In the beginning stages, performance beyond the capability of the student should not be required. Time should be allowed for the student to use each of the pitch instruments to maintain a level attitude. As proficiency develops, small altitude changes should be added to the training. The instructor should not require strict heading control during this phase of training. Since each student is an individual, it is imperative that the instrument instructor carefully evaluate the student and increase training demands in proportion to the student's progress. The instrument instructor must be alert for any problems in instrument scan or interpretation during this early phase of training. Bad habits developed at this time will carry forward in the student's training and may be very difficult to correct at a later date.

ALTITUDE CONTROL

It must be explained to the student that a desired altitude is maintained by establishing a specific pitch attitude on the attitude indicator and trimming the aircraft properly. When the attitude has been established, the vertical velocity indicator, altimeter, and airspeed indicator are scanned to determine if any change is occurring. The altimeter normally is considered the primary pitch instrument during level flight, since it provides the most pertinent altitude information. The supporting pitch instruments include the VSI, airspeed indicator, and attitude indicator. The VSI provides both trend and rate information and immediately reflects initial vertical movement of the aircraft. If a departure from the desired altitude occurs, it is reflected first on the VSI and next on the altimeter. By evaluating the initial rate of movement of these instruments, the student can estimate the amount of pitch change required to restore level flight. The amount of change needed usually is small and requires only a fraction of a bar width of change on the attitude indicator.

When a deviation from the desired altitude occurs, judgment and experience in a particular aircraft dictate the rate of correction. As a guide, the student may be told that to correct small deviations in altitude, the pitch attitude should be adjusted to produce a rate of change which is double the amount of altitude deviation, and power should be used, as necessary. For example, assume an airplane is descending at 300 f.p.m. and is 100 feet below the desired altitude. To correct back to altitude, a climb rate of 200 f.p.m. is selected. An initial pitch adjustment is made to stop the descent and initiate the approximate climb rate. This pitch attitude is maintained on the attitude indicator until the vertical speed stabilizes. A

further pitch adjustment may be necessary to produce the desired climb rate.

HEADING CONTROL

A desired heading is maintained by establishing zero bank on the attitude indicator, with deviations being detected on the heading indicator. The heading indicator is considered the primary bank instrument, since it provides the most pertinent heading information. Supporting bank instruments include the turn coordinator and attitude indicator.

When a deviation occurs, the student should be taught to establish a definite angle of bank on the attitude indicator to produce a suitable rate of turn. When a heading variation of up to 15° occurs, an angle of bank equal to the heading deviation in degrees should be used. If a heading variation exceeds 15°, the degree of bank which produces a standard-rate turn is appropriate; however, the maximum angle of bank should not exceed 30°.

LEVEL TURNS

To enter a level turn, the student should be taught to apply coordinated aileron and rudder pressure in the desired direction of turn. The attitude indicator is used to establish the approximate angle of bank required for a standard-rate turn. The student will soon see that the angle of bank is achieved by aligning the appropriate bank angle mark with the index at the top of the attitude indicator (primary bank).

Since additional lift is required to offset the loss of the vertical lift component in the turn, the pitch attitude must be raised slightly to maintain altitude. As the turn is established, it is necessary to adjust the nose of the miniature airplane so it is slightly above the level flight position on the horizon bar. After the level turn is established, the turn coordinator is primary for bank, and the altimeter is primary for pitch control. The attitude indicator is a supporting instrument for both pitch and bank. The vertical speed indicator is a supporting pitch instrument, and the airspeed indicator is primary power.

The student should also include the heading indicator in the scan to determine progress toward the desired heading. Furthermore, the altimeter should be checked to determine that the adjusted pitch attitude has compensated properly for the loss of vertical lift component and that a constant altitude is being maintained throughout the turn. The principle instrument reference for the roll-out is the attitude indicator. Since a slightly nose-high attitude has been held throughout the turn, elevator or stabilator back pressure is relaxed to prevent altitude gain as the airplane is returned to straight-and-level flight. As the wings-level position is attained, the student should continue the instrument scan. When the airplane stabilizes in the cruise flight configuration, any control pressures required to maintain straight-and-level flight should be trimmed away.

It should be explained to the student that the amount of lead required for roll-out from a turn is approximately one-half the angle of bank. For example, if a standard-rate turn is being made at a bank angle of 15°, the student should be prompted to begin the roll-out approximately eight degrees before reaching the desired heading. For example, the instructor may direct, *"turn the shortest way to the heading of 320°, and report established."* This technique requires the student to mentally calculate reciprocal headings while controlling the airplane. In addition, the student will become accustomed to simplified reporting procedures.

CLIMBS AND DESCENTS

Climbing and descending maneuvers are divided into two general categories — *constant airspeed* or *constant rate*. The constant airspeed maneuver is accomplished by maintaining a constant power indication and varying the pitch attitude, as required, to maintain a specific airspeed. The constant rate maneuver is accomplished by varying both power and

pitch as required to maintain a constant airspeed and vertical velocity. Either type of climb or descent may be performed while maintaining a constant heading or while turning. The student should practice these maneuvers using airspeeds, configurations, and altitudes that correspond to those which will be used in actual instrument flight.

CONSTANT AIRSPEED CLIMBS

To enter a climb from straight-and-level flight at cruise airspeed, the student should increase power to the climb power setting and adjust the miniature airplane on the attitude indicator to approximately two bar widths above the horizon. This pitch adjustment is an average and varies with the aircraft used, the desired airspeed, and rate of climb selected. A smooth, slow power application increases the pitch attitude so that only slight control pressures are needed to effect the desired pitch change. During the transition, the attitude indicator is the primary pitch instrument and a supporting bank instrument. The tachometer or manifold pressure gauge is the primary power instrument. The student also should scan the heading indicator (primary bank) and turn coordinator (supporting bank) to confirm coordinated, straight flight. Since the combination of climb power, torque, and P-factor causes a left-turning tendency, right rudder pressure is required to maintain a constant heading.

CONSTANT AIRSPEED DESCENTS

Constant airspeed descents are performed by the student in a similar procedural manner by using the pitch attitude of the airplane to control airspeed and engine power to control rate of descent. For example, to enter a descent from cruise without a change in airspeed, the power is reduced smoothly to the desired setting and the pitch attitude is reduced slightly so the airspeed remains constant. The degree of pitch change and power reduction will vary according to the particular airplane used for training. Once the power and pitch attitude are established, the airspeed indicator becomes the primary pitch instrument supported by the attitude indicator. Variations in airspeed dictate pitch changes, and small corrections can be made using the attitude indicator.

When the airplane is descending at cruise airspeed, the rate of descent is controlled by small power adjustments. However, when a power change is made, the student must cross-reference the airspeed indicator and attitude indicator to assure the speed remains constant. This procedure becomes important because any change in power also requires a corresponding change in pitch attitude to maintain a constant airspeed during the descent.

RATE CLIMBS

Rate climbs are accomplished by maintaining the desired constant vertical velocity and airspeed. During this maneuver, *pitch attitude control* is used to establish and maintain the *vertical speed*, while *power* is used to *control the airspeed*. From level flight, establish the desired rate of climb by increasing the control back pressure and cross-checking the vertical speed indicator and the attitude indicator. The student should apply power simultaneously in anticipation of the airspeed decrease. As power is increased, the airspeed indicator is primary for pitch control until the vertical speed approaches the desired value; then it becomes primary.

The heading indicator is the primary bank instrument for establishing a straight climb, and the tachometer or manifold pressure gauge is primary for power during the transition. The need for power adjustments after the climb is established can be seen on the airspeed indicator, which is the primary power instrument. A cross-check of the vertical speed indicates the need for subsequent pitch adjustments and a cross-check of the airspeed shows the need for resultant power adjustments.

RATE DESCENTS

Whether a rate descent is made at cruise speed or approach speed, the control procedures are identical. As in a rate climb, power controls airspeed and pitch attitude controls rate.

The student should enter the descent by simultaneously adjusting the nose of the miniature airplane just below the horizon bar on the attitude indicator and reducing power to a predetermined setting. When power is reduced, the aircraft often tends to turn right, so slight left rudder pressure may be required. Once the airplane stabilizes in a constant rate descent, the vertical speed indicator is the primary pitch instrument and the attitude indicator provides supporting pitch information. The airspeed indicator is primary for power control; it is supported by the tachometer or manifold pressure gauge.

LEVELOFF LEADPOINT

The student should understand that the leveloff leadpoint for both climbs and descents is determined using 10 percent of the vertical velocity rate when approaching the desired altitude. For example, if the climb or descent rate is 500 f.p.m., the appropriate leveloff procedure should be initiated when the airplane is 50 feet from the desired altitude. Then, the nose is positioned to a level flight attitude and the altimeter and vertical speed are monitored to maintain level flight. The student then adjusts the power until the desired cruise airspeed is obtained and adjusts trim to relieve control pressure.

PRACTICE MANEUVERS AND PATTERNS

Various maneuvers and patterns combining climbs, descents, turns, and changes in airspeed can be introduced as student proficiency develops. Two examples of such patterns are provided in figure 4-10. Basic VOR navigation may be added as the instrument student progresses in ability.

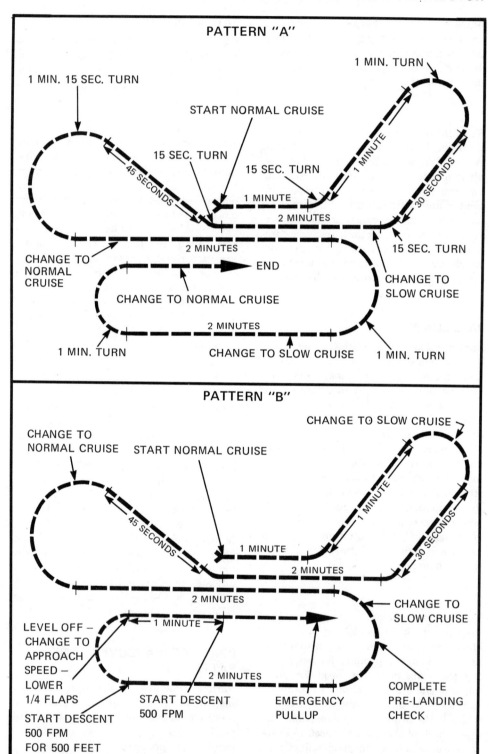

Fig. 4-10. Instrument Training Patterns

As navigation functions are added to the workload, instrument scan problems may arise. The instrument instructor must be careful not to increase the student's workload too rapidly. Since each student will progress at a different pace, the instructor must carefully evaluate each student and adjust the teaching technique accordingly.

AIRSPEED TRANSITIONS

During instrument training, it is necessary for the student to learn to change airspeed while maintaining a constant heading and altitude. Transitions should be introduced and practiced at different airspeeds, varying from cruise to maneuvering at critically slow airspeed. This practice exercise is included for two reasons. First, the maneuver develops coordination during control about all three axes; and second, it prepares the student for the segments of instrument approach procedures in which airspeed must be changed while heading and altitude remain constant. This exercise should be practiced both with and without flaps.

If the proper power settings have been determined for the desired airspeeds, they can be given to the student as an aid in developing speed control techniques. When making large airspeed changes, it is frequently desirable to undershoot the power setting, allowing the aircraft to decelerate more rapidly. As the airspeed approaches the desired speed, the power should be set to stop the deceleration. As soon as the desired airspeed is attained, the airplane should be trimmed for the new attitude.

During an airspeed *reduction*, the altimeter and vertical velocity indicator are used to establish the correct pitch attitude and the airspeed indicator is used to determine the power setting. It must be remembered that the altimeter does not indicate the *amount* of pitch, but only if the pitch attitude is correct. In this sense, the airspeed indicator does not indicate how must power is being used, but whether the correct power setting is being used. As the speed dissipates, the heading indicator

and the turn coordinator should be referenced to determine that the desired heading is being maintained.

Heading control usually is the most common problem for a student performing this maneuver. The student should maintain a proper scan of the instruments and not become so engrossed with the pitch instruments that the heading indicator is ignored.

Another common student difficulty is the tendency to use the pressure instruments for pitch control, which results in overcontrolling the aircraft. More accurate pitch control results if the student uses the attitude indicator as the principal pitch control instrument and the pressure instruments for confirmation.

STALLS

Although imminent and full stalls are not required for instrument pilot certification, they are valuable training maneuvers for teaching aircraft control. In addition, the practice of stalls improves instrument scan techniques and also builds student confidence.

POWER OFF

During practice of power-off stalls, it is important for the instructor to relate the maneuver to actual in-flight situations. The student should establish the aircraft in the configuration used on an instrument approach, including selection of the necessary power settings, extension of approach flaps and, if appropriate, extension of the landing gear. The student should monitor the altimeter and increase the pitch attitude gradually so there is no loss or gain in altitude. As the pitch is increased, the attitude becomes too great for the available power, causing the airspeed to dissipate gradually to the stalling speed.

Recovery from the stall should be initiated by a smooth, immediate pitch reduction to level flight or a slightly nose-low attitude followed by the smooth application of full power. The student should be taught to maintain this attitude until the best angle-of-climb airspeed is obtained. When this speed is established, the airplane nose should be raised to the climb attitude. If the aircraft is in turning flight or it yaws upon entering the stall, coordinated aileron and rudder pressure should be used to maintain the desired heading. Since the conditions which normally cause a stall are encountered close to the ground, the student should recognize the need for establishing a slight climb as soon after recovery as possible to facilitate a minimum loss of altitude during the recovery.

POWER ON

The student should understand that the pitching of the nose during a power-on stall is steeper and more rapid than that experienced during a power-off stall. After the power is established, the nose should be raised to a pitch attitude which the airplane cannot maintain. Usually, this is three to four bar widths above the horizon on the attitude indicator. In addition, the airplane is slightly more difficult to control during stalls in the climbout configuration because there may be a tendency for a rapid roll rate to develop. There is also a strong left-turning tendency due to the increased torque and P-factor. However, the stabilator or elevator remains effective throughout the stall due to the propeller slipstream. This causes a higher pitch attitude before the stall, but it also aids in a more positive stall recovery.

On the first recovery, the student is prone to force the nose of the aircraft too far below the horizon. Once in a nose-low position, it is advisable for the instructor to help the student gently ease the aircraft back to level flight, explaining that harsh control movement may cause excessive wing loading and/or a secondary stall.

UNUSUAL ATTITUDE RECOVERY

Spatial disorientation, turbulence, lapse of attention, or abnormal trim conditions can cause a pilot to enter an unusual or critical flight attitude. Although such cases are rare, a student must know how to recover from an unusual attitude during instrument flight.

To enter an unusual attitude, the instructor flies the airplane through various maneuvers calculated to induce spatial disorientation. The student's eyes should be closed and the hands and feet removed from the controls. At a critical point, the instructor tells the student to take control and recover the aircraft to the original heading and altitude.

NOSE-HIGH ATTITUDE

The instrument indications of a typical nose-high unusual attitude are a very high pitch attitude, decreasing airspeed and, normally, a high rate of climb. Before initiating a correction, the student should cross check the instruments to confirm the reliability of the attitude indicator. The initial objective for recovery from this attitude is to prevent a stall. Therefore, the student should simultaneously increase power, decrease pitch (reducing angle of attack), and roll the wings level.

NOSE-LOW ATTITUDE

The indications of a nose-low unusual attitude are a very low pitch attitude, increasing airspeed, rapid loss of altitude, and a high rate of descent. The primary objective of a nose-low unusual attitude recovery is to avoid a critically high airspeed and load factor. To accomplish this, the student should simultaneously reduce power, roll the wings level, increase pitch attitude to stop acceleration, and gently raise the nose to the level flight attitude. If the student raises the nose before rolling the wings level, the increased load factor may result in an accelerated stall, a spin, or a force exceeding the aircraft design limit load factor.

INSTRUCTING PARTIAL PANEL

Practice of partial panel procedures is widely accepted as a necessary ingredient of instrument training. In a broad sense, partial panel refers to the loss of any of the six basic flight instruments. During training, it usually means controlling the airplane without reference to the attitude or heading indicators. Loss of these two instruments could realistically result from the failure of the vacuum system in typical light, general aviation airplanes.

In actual IFR flight, the failure of the attitude and heading indicator is considered a semi-emergency situation because the pilot may not be able to comply immediately and accurately with ATC clearances. In addition to preparing the instrument student for an actual emergency, partial panel training has several other advantages. Students who receive partial panel training during basic instrument flight instruction develop more accurate full panel procedures and, generally, progress faster in training. They are quicker to recognize instrument failure in an actual situation and are conditioned to the intense concentration that partial panel flying requires.

Many instructors prefer to introduce partial panel maneuvers after initial practice of all basic maneuvers using full panel references. This allows more diversity in training during the basic attitude instrument phase which otherwise may seem somewhat boring to instructors and students. A given lesson may begin with full panel but conclude with partial panel when the instructor covers the heading and attitude indicators to simulate failure of the vacuum system. Partial panel should include the same basic maneuvers used during full panel practice. Many experienced instructors feel that one hour of partial panel is as effective as two hours of full panel in terms of student progress, as shown in figure 4-11.

Compass turns on partial panel require the student to have a precise knowledge of the characteristics of the magnetic compass. The student should be directed to make successive turns (both left and right) to the cardinal compass headings.

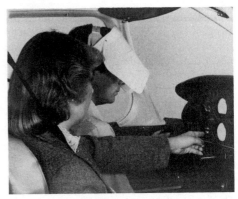

Fig. 4-11. Partial Panel Practice

This should be accomplished without timing in smooth air conditions. The amount of lead or lag during roll-out for northerly or southerly headings will be approximately equal to the latitude of the training location plus five degrees for the time required to establish or stop a standard-rate turn.

Timed turns also are valuable procedures for partial panel practice. Typically, the instructor directs, *"Turn the shortest way from your present heading to the heading of 250°"*. This requires the student to check the magnetic compass, determine the number of degrees in the turn, and compute the time required to complete the turn at standard rate (three degrees per second).

For increased accuracy, the turn should be started when the secondhand on the aircraft clock passes one of the cardinal numbers. At the end of the predetermined number of seconds required to make the desired heading change, the student should begin to roll out of the turn. Roll rate should be constant for both roll in and roll out.

After the aircraft is stabilized in level flight, the student checks the magnetic compass to determine whether the turn was completed on the desired heading. Any necessary corrections are accomplished by using the same procedure, then rechecking the magnetic compass after it has stabilized.

Timed turns have realistic application to the failure of the heading indicator in actual IFR conditions. They also condition the student to perform mental calculations while flying the airplane with a minimum of instrument references. After a few hours practice, instrument students should be able to turn rapidly and precisely to any designated heading in smooth air conditions.

NAVIGATION

During the latter phase of basic attitude instrument flying, most instructors prefer to introduce VOR and NDB navigation procedures. This is basically a full panel exercise, but some partial panel practice also can be beneficial. The student will have at least a basic familiarity with VOR and NDB navigation from past experience in VFR operations. However, the instructor should provide a detailed briefing on the distinctive procedures used for IFR flight prior to the first lesson on navigation.

Navigation exercises obviously require the student to include additional instruments in the cross-check. The instructor should emphasize that the VOR and ADF indicators are not to be used as the principle heading control instruments. The student should refer to them only to confirm position and determine the need for heading corrections. Beginning students have a tendency to use the course deviation indicator and the ADF bearing indicator as though they were heading indicators. This tendency must be broken early in training or it will interfere with navigation procedures throughout the training program.

BASIC VOR NAVIGATION

Several methods are used to position an aircraft on a specific VOR radial to obtain course guidance. All methods use magnetic headings which intercept the course or airway at a specific angle. The student should understand that the intercept angle is the number of degrees formed by the intersection of the projected aircraft heading and the desired course.

INTERCEPTING VOR RADIALS

In figure 4-12 for example, an aircraft departing the airport may select a heading of 300° to intercept V-14 westbound. This procedure results in a 30° outbound intercept. At course interception, the CDI centers, and the aircraft should be turned to a heading of 270°. Depending on airport location and local terrain, a different intercept angle may be used. However, 30° is a popular angle for enroute operations. This intercept allows the student time to turn, upon interception of the selected radial, without "overshooting." In cases where large intercept angles such as 60° or 90°, are used, it becomes necessary to "*lead*" the turn to the selected radial. *Lead* is used when in close proximity to a VOR and prevents overshooting. The degree of lead varies with the groundspeed of the aircraft and the distance from the navaid, as well as the intercept angle.

A situation in which a student plans to fly eastbound on V-14 after departing from the airport is depicted in figure 4-13. An intercept heading of 360° could be used, resulting in a 90° inbound intercept. The turn on course (090°) requires a lead factor of approximately five degrees. However, an initial bearing selection of 085° could also be used to

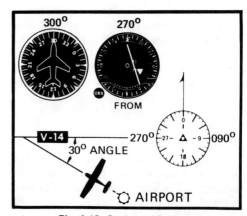

Fig. 4-12. Outbound Intercept

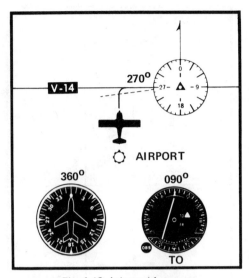

Fig. 4-13. Inbound Intercept

indicate when the inbound turn should begin. After starting the turn, the course selector should be changed to 090°. The actual roll-out heading will depend on the CDI deflection and rate of movement when approaching a heading of 090°. A wide range of *intercept angles* may be suitable; however, the angle of intercept obviously must not exceed 90°.

TIME AND DISTANCE

A popular training maneuver, used to determine time and distance to a navaid, is illustrated in figure 4-14. While on a near perpendicular course to a VOR radial, the time required to cross

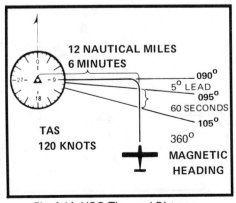

Fig. 4-14. VOR Time and Distance

a predetermined *radial span* can be used to determine the time and distance to the navaid.

In the example shown, the student plans a 90° intercept for the 090° radial. The passage if a 10° change (105° to 095°), while on a heading of 360°, requires 60 seconds. Radial spans of 5°, 15°, or 20° may also be used; however, the 10° radial span facilitates division. By applying the following formula, the approximate time to the station can be determined:

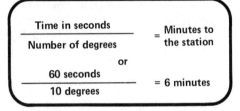

Therefore, if the time check is completed at 1215Z, the airplane will arrive over the VOR at approximately 1221Z. At a speed of 120 knots TAS (approximately two nautical miles per minute depending upon wind), the aircraft will be 12 nautical miles east of the VOR at the completion of the time check (2 n.m./min. x 6 min. = 12 n.m.).

REVERSE SENSING AND RECIPROCAL BEARINGS

During initial training, many students will periodically set a reciprocal course on the VOR indicator and consequently track off course. The best way to avoid reciprocal tuning and reverse sensing is to require the student to mentally verify that the *desired magnetic course, actual magnetic heading,* and *course selection* are in general agreement. When this is accomplished, a TO indication is always displayed approaching a VOR, and a FROM indication will occur when flying outbound from a VOR.

BASIC ADF NAVIGATION

Tracking to an ADF facility in a *no-wind condition* simply requires turning the aircraft to a heading which results in a zero degree relative bearing. The student then maintains a constant heading which results in a straight course to a navaid.

The same is true when tracking away from the NDB. In this instance, however, the relative bearing is 180. ADF tracking in a *crosswind* is somewhat more involved, since unlike VOR, ADF does not provide an automatic wind correction angle. In order to track accurately in a crosswind, the student must learn to apply the ADF formula. When on course, the *wind correction angle* should be exactly *equal to the number of degrees a station bears to the left or right* of the nose of the aircraft (MH + RB = MB to station).

ADF INTERCEPTS

During initial practice of ADF intercepts, it may be beneficial for the student to parallel the desired course before beginning the intercept. For example, in figure 4-15, the student determines the airplane's position to be on the 195° bearing from an NDB. If the desired inbound course is the 180° bearing, a turn is made to a heading of 360° to parallel the course. From this position, it is easier to visualize the intercept procedure. The selection of a 30° intercept heading results in the intersection of two parallel lines at equal angles. This geometric principle is involved in all tracking and intercept procedures using ADF navigation.

After proficiency is obtained, paralleling can be discontinued and more rapid intercept procedures can be used. For example, in figure 4-16, assume an aircraft is located on the 250° bearing from an NDB and the student is directed to track inbound on the 270° bearing from the station. A heading of 060° will produce a 30° intercept angle, therefore, the student will intercept the 270° bearing when the relative bearing is 30°. The aircraft should then be turned to a heading of 090° to track inbound to the station.

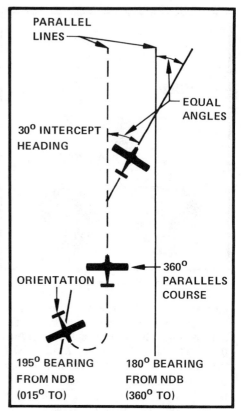

Fig. 4-15. Paralleling the Course

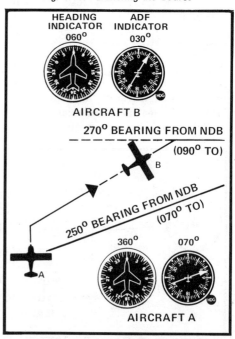

Fig. 4-16. Inbound Intercept

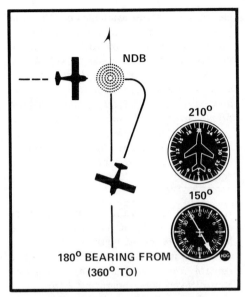

Fig. 4-17. Outbound Intercept

of the aircraft approaches the intercept angle (30°). In this example, when a relative bearing of 150° is approached, the student should turn the aircraft to a heading of 180° and track outbound. A continuous 180° relative bearing indicates that the aircraft is maintaining course.

The student should be aware that *reliable ADF navigation* is dependent upon the *accuracy* of the *heading indicator*. If the heading indicator has precessed 10°, the tracking procedures and intercepts will be in error 10°. Although this will not be very significant during inbound tracking, during outbound tracking over large distances, the error is multiplied. A pilot can be miles off course and yet be completely unaware of the predicament, unless agreement between the magnetic compass and the heading indicator is confirmed.

TIME AND DISTANCE

Time and distance checks are similar to those used with VOR. The ADF procedure is somewhat easier, since the consecutive course selector changes required for VOR are not needed with ADF. In figure 4-18, assume the student performs

Assume that the student is then directed to depart the NDB on the 180° bearing, as shown in figure 4-17. As in VOR operations, various intercept angles may be used. In this case, a 30° intercept heading of 210° is chosen.

The turn on course should be started when the station's bearing from the tail

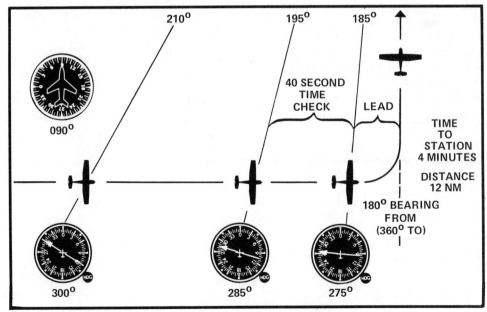

Fig. 4-18. ADF Time and Distance

an ADF orientation and determines the airplane location to be on the 210° bearing from an NDB. A 90° inbound intercept is planned for the 180° bearing with a time and distance check enroute.

The first step is to turn the aircraft to a heading of 90°, which gives a 90° intercept to the 180° bearing. Since a relative bearing of 270° will indicate interception of the 180° bearing, the time check should begin 15° before interception, or at a *relative bearing* of 285°. The check should be completed at 275° *relative bearing*, leaving 5° of lead for the turn inbound. The same formula as used in VOR applies; that is:

$$\frac{\text{Time in Seconds}}{\text{Number of Degrees}} = \frac{\text{Minutes to}}{\text{the station}}$$

If the check requires 40 seconds, the time to the station is approximately four minutes in a no-wind condition. At 180 knots TAS (3 nautical miles per minute), the distance to the station is approximately 12 nautical miles, depending on wind conditions.

Time and distance checks are useful training maneuvers, because good position visualization and mental calculations are required while flying the airplane through a navigation exercise. They provide a logical transition to holding pattern and approach procedures training.

SIMULATED CLEARANCES

During the basic navigation phase, the instructor should provide simulated ATC clearances to the student on a regular basis. For example, a simulated IFR departure clearance from an uncontrolled airport might be, *"Cherokee 2014Q, cleared to the practice area via heading 230°, climb and maintain 2,500, turn left after departure, squawk 1200 just before departure."* This type of simulation will provide practice in clearance copying and readback and will prepare the student for actual ATC communications later in the course.

As the student gains familiarity with clearance procedures, more complex clearances may be issued by the instructor. For example, it is easy to simulate ground control, tower and approach/departure control even when operating from uncontrolled airports. Assigning mock frequencies for each ATC function will require the student to actually switch frequencies on a secondary transceiver during handoff from one simulated facility to another.

In addition, ATC communications and clearances should be practiced during basic navigation exercises whenever possible. For example, *"Cherokee 2014Q, state your position from the Aberdeen VORTAC and say your altitude."* After the student performs the VOR orientation and reports, a clearance may be issued such as, *"Cherokee 2014Q, cleared to the Aberdeen VORTAC from present position via the 180° radial, maintain 2,500, time and distance check enroute, report inbound on the 180° radial with your Aberdeen estimate."* Although ATC would not issue such a clearance, the student will benefit from the exercise.

SECTION B—INSTRUMENT PROCEDURES INSTRUCTION

This section covers instrument procedures typically used in IFR operations. These procedures include takeoff, departure, holding patterns, approaches, and cross-country flights. This phase of training generally is considered the most interesting, but instructors should remember that successful training in advanced instrument flight is totally dependent on student proficiency gained in basic attitude instrument flight. In addition, advanced training will require more planning and preparation for both the student and instructor.

Instrument procedures also require the student to have a solid academic knowledge of IFR operations. To ensure rapid progress in this phase of training, the instructor must monitor the student's progress in the ground training segment of the course. Prior to advanced instrument flight training, the student should have completed a study of IFR regulations, instrument flight charts, clearances, and ATC procedures. Accomplishing this will provide a framework for IFR procedures, and the workload of both instructor and student will be greatly reduced.

PLANNING

Accurate and thorough planning is necessary for any flight; especially a flight under instrument conditions. Monitoring the student's weather briefing is a good way to see if the right questions are being asked. The instrument flight instructor can reinforce the habit of careful planning by reviewing each lesson with the instrument student as a part of the preflight preparations. This will help ensure that both instructor and student understand what is to be accomplished in each lesson. In addition, performance standards should be set and clearly understood to permit an objective evaluation of the lesson after its completion.

One of the most common beginning instrument student errors is the lack of planning. The instrument instructor must guide the student in this advanced planning and not expect it as a matter of course. For example, if holding patterns are part of the lesson, the student should be requested to mentally rehearse each phase of the procedure. The instructor should not hesitate to assign pertinent sections of student study materials in preparation for specific flight lessons.

In addition to checking the aircraft and engine logbooks for the periodic and 100-hour inspections, the instrument student should be concerned with recent VOR accuracy checks and static system checks. These checks are entered in the aircraft logbook. Airborne VOR equipment used on an IFR flight either must be maintained, checked, and inspected under an approved procedure, or it must have been operationally checked within the preceding thirty days. The person making the accuracy check must enter the date, location, bearing error, and signature in the aircraft log. The student also should be conditioned to check the operational status of the communications radios and equipment. This includes headsets, push-to-talk buttons, and headsets or boom microphones. Many instructors carry an extra microphone as insurance against failure.

In addition, each static pressure system and each altimeter must have been tested and inspected in accordance with FAR Part 91.171. These checks must have been accomplished within the preceding 24 calendar months for every airplane operated under IFR in controlled airspace. A record of the testing and inspection normally is kept in the aircraft logbook.

INSTRUMENT CHECKS

The student should understand that the preflight check of the gyro-operated instruments and their power sources takes on additional importance when planning a flight under instrument conditions. It is far better to find an inoperative instrument or system during the preflight than it is to cope with it after departure. Before the master switch is turned ON or the engine is started, the student should make sure the instruments that have power indicators are displaying OFF indications. These indicators are the first sign that an instrument has failed. The inclinometer should be full of fluid, with the ball resting at its lowest point.

When the master switch is turned on, there should not be any abnormal noises, such as grinding sounds, that would indicate an impending failure. After the engine is started, the student should listen to the vacuum-driven gyros. The gyros normally reach full operating speed in approximately five minutes. During this time, it is common to see some vibration in the indications. When the gyros have stabilized, the miniature airplane in the turn coordinator and the horizon bar in the attitude indicator should be level while the airplane is stopped or taxiing straight ahead. During turns, the turn coordinator (or turn-and-slip indicator) and heading indicator should display a turn in the correct direction. The ball in the inclinometer should swing to the outside of the turn. The attitude indicator should not tilt more than five degrees during normal turns. The magnetic compass should be checked against known headings on the airport and the heading indicator aligned with the compass. The heading indicator is rechecked prior to takeoff to ensure it has not precessed significantly. A precession error of no more than 3° in 15 minutes is acceptable for normal operations. Other instruments which should be checked are the engine instruments, clock, and flap and trim indicators.

IFR DEPARTURES

Instrument takeoffs and departures can be accomplished smoothly and efficiently when the student is familiar with ATC procedures. This keeps radio communications to a minimum and allows the student to concentrate on the transition to instru-

ment references. The navigation workload varies with the procedure used. It may be as simple as holding an assigned heading while monitoring navigation aids or adhering to a detailed standard instrument departure (SID) procedure. At some airports, the student may perform the departure procedure without SIDs or radar assistance. In any event, there are several details relating to takeoffs and departures that the student must learn.

TAKEOFF MINIMUMS

FAR Part 97 prescribes standard instrument approach procedures for instrument letdowns to airports in the United States. It also prescribes the weather minimums that apply to takeoff for commercial operators from most airports which have a published instrument approach procedure. If takeoff minimums for a particular airport are not prescribed in Part 97, the following standard minimums apply: one mile visibility for single-engine and twin-engine aircraft, and one-half statute mile visibility for aircraft with more than two engines. The so-called standard takeoff minimums are specified in FAR Part 91.175(f).

The student should be advised that, although these standard requirements do not specifically apply to private aircraft operations conducted under Part 91, good judgment dictates compliance. The prudence of an IFR takeoff in a light aircraft when weather conditions are known to be below commercial takeoff minimums is, to say the least, questionable.

VISIBILITY AND RVR

Normally, takeoff minimums are expressed in terms of prevailing visibility or runway visual range (RVR). Prevailing visibility is the greatest distance an observer can see throughout half of the horizon. RVR is used when transmissometer equipment is installed. RVR values are always reported with respect to a particular runway; they may not be applied generally to other adjacent runways or to the entire airport.

TAKEOFF MINIMUMS (JEPPESEN FORMAT)

If an instrument approach procedure is prescribed for an airport, takeoff minimums *also* will be shown on Jeppesen approach charts. These charts show take-off minimums for each airport, whether they are standard or nonstandard. In addition, Jeppesen charts show lower-than-standard takeoff minimums which may be authorized for certain commercial operators. Runways where reduced visibility takeoffs are permitted have to be equipped with certain lighting or marking aids to make these operations safe. Although legal, it is unwise for a Part 91 operator to use these lower-than-standard minimums.

In other cases, greater-than-standard take-off minimums may be specified because of terrain or other obstructions. Ceilings, as well as visibilities, are specified where a minimum ceiling is required in order to avoid obstructions during departures. Where necessary, detailed IFR *departure procedures*, including items such as routes and minimum altitudes, are provided. For example, the takeoff minimums for Midland Airpark, as shown in figure 4-19 (item 1), specify an IFR departure procedure for runway 11. A 700-foot ceiling and one-mile visibility are required for this runway unless a minimum climb gradient of 300 feet per nautical mile can be maintained to 3,500 feet MSL. Billings, Montana, (item 2) is another example where obstructions (in this case mountains) require a minimum ceiling and greater-than-normal visibility for takeoff from runway 9L. Very detailed departure instructions are given at the bottom of the excerpt.

Fig. 4-19. IFR Takeoff Minimums and Departure Procedures (Jeppesen Format)

TAKEOFF MINIMUMS (NOS FORMAT)

When takeoff minimums are not specified on NOS charts, the standard minimums (one mile or one-half mile) apply. However, the symbol ▼ in the information box at the bottom of the chart indicates that special minimums and/or procedures have been established for the airport concerned. Therefore, the pilot must consult a separate NOS tabulation, entitled " ▼ IFR Takeoff Minimums and Departure Procedures," which is contained in the approach chart binders. Figure 4-20 shows an excerpt of the tabulation. The absence of an airport in the tabulation means only standard takeoff minimums apply. Regardless of the charts used (Jeppesen or NOS), the instructor must ensure the student studies the legends carefully.

DEPARTURE TRAINING

The instrument instructor must prepare the student for a variety of departure procedures during the training program. They include radar, nonradar, and standard instrument departures (SIDs). To accomplish this, simulated IFR departures, complete with clearances and communications, should become standard practice during instrument procedure instruction. The instructor should require the student to lower the hood within 100 or 200 feet AGL following takeoff. This provides a realistic transition to instrument references, and is similar to what would be experienced during an actual IFR departure. It also conditions the student for the immediate concentration which is required for the transition. The instructor can then simulate radar departure control and provide vectors to the practice area, or a nearby navigation aid, to begin the actual lesson. A "nonradar" departure clearance can be issued to accomplish the same purpose. In other cases, a SID may be assigned. The procedure can be a published version or one designed by the instructor for use at the training airport.

HOLDING PATTERNS

Holding patterns and entries are added to the training as soon as the student has

---

INSTRUMENT APPROACH PROCEDURES (CHARTS)
▼IFR TAKE-OFF MINIMUMS AND DEPARTURE PROCEDURES

Civil Airports and Selected Military Airports

CIVIL USERS: FAR 91 prescribes take-off rules and establishes take-off minimums as follows:
(1) Aircraft having two engines or less – one statute mile. (2) Aircraft having more than two engines – one-half statute mile.

MILITARY USERS: Special IFR departure procedures, not published as Standard Instrument Departure (SIDs), and civil take-off minima are included below and are established to assist pilots in obstruction avoidance. Refer to appropriate service directives for take-off minimums.

Airports with IFR take-off minimums other than standard are listed below. Departure procedures and/or ceiling visibility minimums are established to assist pilots conducting IFR flight in avoiding obstructions during climb to the minimum enroute altitude. Take-off minimums and departures apply to all runways unless otherwise specified. Altitudes, unless otherwise indicated, are minimum altitudes in feet MSL.

NAME	TAKE-OFF MINIMUMS	NAME	TAKE-OFF MINIMUMS

MIDLAND AIR PARK, TX
                    Rwy 11, 700-1*
Rwy 16, not authorized.
*or standard with minimum climb of 300' per NM to 3500.
IFR DEPARTURE PROCEDURE: Rwy 11, climb runway heading to 3500, then climb on course.
		BILLINGS LOGAN INTL, MT
		   Rwy 9L, 800-2 or standard with minimum climb of 270' per NM to 4500. Rwys 9R, 27L, NA.
		IFR DEPARTURE PROCEDURE: NE bound V2, E bound V247, SE bound V19/V86, W bound V2/V86, V247, NW bound V19/V187, climb on course. All others climb direct BIL VORTAC. Continue climb in holding pattern (W, RT, 070° inbound) to cross BIL VORTAC at or above: S bound V85 4600. All others MCA or MEA for direction of flight.	

Fig. 4-20. IFR Takeoff Minimums and Departure Procedures (NOS Format)

demonstrated a thorough understanding of the VOR and is capable of accurate VOR intercepts and tracking. During this phase of training, the instructor has a good chance to observe the student's ability to maintain heading and altitude while performing standard rate turns, and handling the additional problems of clearances, radio communications, and navigation. To most effectively teach holding patterns, the instrument instructor must be well versed in entry procedures. The following discussion is included as a brief refresher on the subject.

ENTRIES

Standard holding pattern entry procedures are designed to reduce maneuvering and conserve airspace during the initial phase of the holding procedure. Three types of entry procedures are used.

1. Direct
2. Teardrop
3. Parallel

The direct entry applies to the sectors 70° left through 110° right of the holding course, as viewed from the holding fix. The teardrop sector covers 70° of azimuth, and the parallel sector includes 110°. Aircraft magnetic *heading* upon arrival at the fix determines the type of entry to use. Plus or minus five degrees in heading is considered within good operating limits for determining the correct entry. Therefore, flights approaching a fix on sector boundaries may use either of two procedures in most cases.

DIRECT ENTRY

The direct entry procedure is the simplest pattern entry. It is also the one most often encountered because it can be applied throughout 180° of azimuth in relation to the holding fix. When using a direct entry to a standard pattern, the student simply crosses the fix, turns directly to the outbound heading, and flys the pattern. Direct entry procedures are depicted in figure 4-21. Notice in the

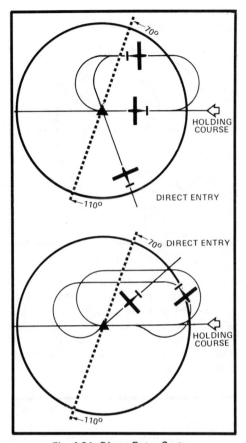

Fig. 4-21. Direct Entry Sector

two examples that the aircraft may not be perfectly established on the holding course until the second circuit of the holding pattern.

TEARDROP ENTRY

The teardrop entry is the next simplest procedure. After crossing the fix, the student should turn to an outbound heading which diverges approximately 30° toward the holding side of the course, as shown in figure 4-22. After holding the heading for 1 minute (or 1½ minutes, if appropriate), the aircraft should be turned to intercept the holding course inbound and return to the fix.

PARALLEL ENTRY

The parallel entry procedure is used throughout 110° of azimuth and involves

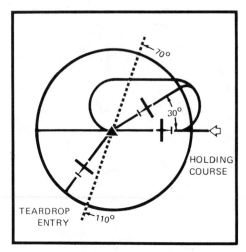

Fig. 4-22. Teardrop Entry Sector

paralleling the holding course outbound on the nonholding side. After one minute, a turn is made to intercept the inbound course. It may be difficult to actually intercept the holding course before recrossing the fix, as illustrated in figure 4-23.

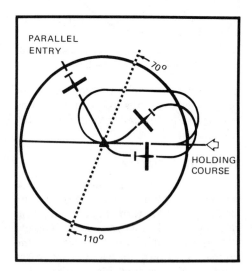

Fig. 4-23. Parallel Entry Sector

VISUALIZING ENTRY PROCEDURES

Instructors and students have improvised various methods for visualization of entry procedures prior to arrival over the fix. They include drawing holding patterns on aeronautical charts, using the

wind side of a flight computer, employing the aircraft heading indicator, referring to a holding pattern indicator, referring to a holding pattern inscription on a plotter, or using specially designed pattern entry computers. The value of any of these methods is directly proportional to how well they actually help the student to *mentally visualize* the holding pattern and the appropriate entry procedure.

An alternate method which does not require separate equipment may be the most advantageous. One of these methods requires the student to visualize arrival at the fix. If the holding course is behind the aircraft upon arrival at the fix, a *direct entry* is appropriate. When the holding course is *ahead* of and to the *right* of the aircraft, a *teardrop entry* may be used. A *parallel entry* is appropriate if the holding course is ahead and to the *left* of the aircraft as it crosses the fix.

This method of visualization and the various methods of entry may also be applied to nonstandard holding patterns. The direct entry sector is still behind the aircraft; however, holding courses ahead and to the right of the aircraft require a parallel entry. Conversely, holding courses ahead and to the left of the aircraft require a teardrop entry.

OUTBOUND AND INBOUND TIMING

After entry into a holding pattern, the initial outbound leg should be flown for 1 or 1½ minutes, whichever is appropriate for the altitude. The timing for subsequent outbound legs should be adjusted to achieve *proper inbound timing* (1 minute or 1½ minutes). This procedure is required because of the effect of wind which may cause significant differences between the inbound and outbound leg groundspeed.

If the inbound leg of a holding pattern requires 45 seconds instead of one minute, the outbound leg should be lengthened. The normal procedure is to length-en the outbound leg by the same amount that the wind has shortened the inbound leg (in this case, 15 seconds). Although this procedure results in slightly unequal time for the two legs, it will suffice unless winds are extremely high in relation to the true airspeed of the aircraft. Students should continue to check the inbound timing and make outbound adjustments as necessary.

CROSSWIND CORRECTION

Crosswind conditions produce the necessity for pronounced heading corrections. The student must be aware that crosswind corrections are made by initiating heading corrections on the inbound and outbound legs of the holding patterns; corrections are not used during the turns. During inbound tracking, the student should determine the wind correction angle necessary to maintain course from the navigation indicators. However, as shown in figure 4-24, item 1, if the same amount of wind correction is applied during the outbound leg, the aircraft will overshoot the course following the turn to the inbound leg. Item 2 shows the result of *tripling* the

WCA when flying the outbound leg. Failure to apply this technique may result in penetration of the airspace on the nonholding side. In addition, it may be difficult to become established on the inbound course before reaching the fix, resulting in more pattern irregularity and possibly contributing to disorientation.

ATC HOLDING INSTRUCTIONS

If the holding pattern assigned by ATC is depicted on the appropriate aeronautical charts, the pilot is expected to hold according to the procedure depicted, unless advised otherwise by ATC. An ATC clearance assigning a holding pattern which is not depicted on the appropriate chart will contain the information shown in figure 4-25.

Instrument pilots are to hold at the last assigned altitude unless a new altitude is specifically given. When controllers anticipate a delay at a clearance limit or fix, they usually will issue a holding clearance at least five minutes prior to the aircraft's arrival at that position. In addition, pilots are well advised not to accept a holding clearance without an *expect*

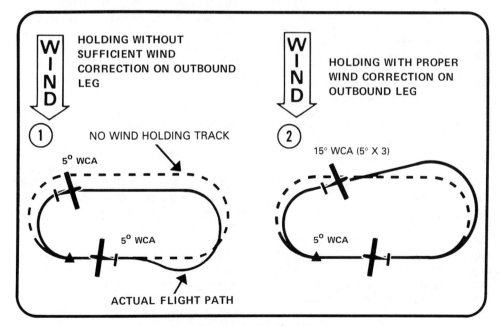

HOLDING WITHOUT SUFFICIENT WIND CORRECTION ON OUTBOUND LEG

WIND

① NO WIND HOLDING TRACK

5° WCA

5° WCA

ACTUAL FLIGHT PATH

HOLDING WITH PROPER WIND CORRECTION ON OUTBOUND LEG

WIND

② 15° WCA (5° X 3)

5° WCA

Fig. 4-24. Holding in a Crosswind

ELEMENTS OF HOLDING CLEARANCE	EXAMPLE OF HOLDING CLEARANCE
1. The direction to hold from the holding fix	*". . . Hold west*
2. Holding fix	*of the Greenwood Intersection*
3. The specified radial, course, magnetic bearing, airway number, or jet route	*on V-8,*
4. The outbound leg length in minutes or nautical miles when DME is used	*five-mile legs,*
5. Nonstandard pattern if used	*left turns.*
6. Expect further clearance time	*Expect further clearance at 15."*

Fig. 4-25. ATC Holding Instructions

further clearance (EFC) time because of the possibility of two-way communications failure. Without an expect further clearance time, the pilot would not have a specific time to depart the holding fix.

HOLDING PATTERN COMMUNICATION REPORTS

Pilots should provide *corrected estimates of arrival* at assigned holding fixes whenever it becomes apparent that the previous estimate was in error by *three or more minutes.* Pilots also should report *entering,* as well as *leaving,* the holding point or fix. Revised estimates, pertaining to the holding pattern should be made only when not in radar contact. However, *changes in assigned altitude* while holding should be reported *at all times.* After they depart a holding pattern, pilots are expected to resume the original true airspeed shown on the flight plan.

APPROACH PROCEDURES

The execution of instrument approaches requires a high level of instrument proficiency for both instructor and student. Aircraft control, navigation, chart interpretation, ATC procedures, and radio communication all are involved during instrument approaches. The function of

instrument flight training is to teach the student smooth, accurate coordination of these operations and procedures.

It is not intended that the sequence of approaches presented in this discussion be the one used during training. In many cases, the instrument facilities at the training airport, or one nearby, will be the determining factor. The instructor should introduce the various approach procedures as rapidly as student progress permits, rather than emphasizing one particular procedure until flight test proficiency is attained. In this manner, several procedures may be practical during a given lesson to provide variety and keep student interest high.

VOR APPROACHES

The planning for an instrument approach must begin prior to flight. The preflight discussion should center around the same approach that will actually be performed during the flight lesson. The appropriate charts should be laid out and the route of flight traced. Each step of the approach should be covered, including radio frequencies and communications. Questions concerning procedural sequences and altitudes should be answered in detail.

VOR approaches are two general types— those where the VOR facility is located on the airport, and those where the facility is located beyond the airport boundaries. The VOR approach which will be discussed in this section utilizes an off-airport facility.

The training flight, illustrated in figure 4-26, is enroute to Sheridan County Airport, Sheridan, Wyoming via V-19 at an assigned altitude of 9,000 feet. Since this route crosses an ARTCC boundary, the student should be advised to expect a handoff from Denver Center to Salt Lake City Center after passing Crazy Woman VORTAC. In anticipation of the handoff, the student should locate the Salt Lake City Center frequency on the enroute chart and set it in the number two communications radio.

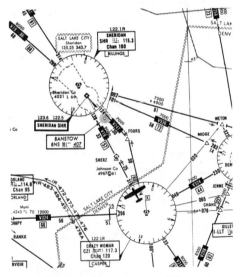

Fig. 4-26. Enroute Segment

APPROACH CHART REVIEW

The student also should review the VOR runway 13 approach chart (NOS format) to Sheridan County that is shown in figure 4-27. Since the Sheridan VORTAC is part of the enroute structure, an approach clearance will allow the flight to cross the facility and proceed outbound for the procedure turn while descending as rapidly as practical to the minimum procedure turn altitude of 7,000 feet MSL. The student should be reminded that the assigned

altitude on V-19 is the MEA, so descent cannot be initiated until the VORTAC is crossed. Additional factors the student should note during the approach chart review are the 5,600-foot minimum altitude over the VORTAC inbound, time to MAP, as well as the MDA, and missed approach procedure.

The profile and landing minimums sections of this approach chart reveal that DME equipment can be used for the procedure although it is not required. Many VOR procedures utilize supplemental DME fixes, but absence of the notation "DME" in the chart heading means the use of DME equipment is not mandatory for the approach. However, the use of DME may permit a lower MDA, which is the case for the Sheridan approach. These details should be brought to the student's attention.

FLYING THE APPROACH

The center handoff, in this situation, will occur soon after the flight crosses the Crazy Woman VORTAC. The student should be reminded to reset the altimeter, if necessary. Following the handoff to Salt Lake City Center, an approach clearance will be issued and the student will be advised to contact Sheridan Radio on 123.60 MHz for an airport advisory. This action should be taken promptly so the student can determine from the wind conditions whether a straight-in landing to runway 13 can be made or a circling approach will be required.

Prior to reaching the fix, a speed reduction should be accomplished. As the flight crosses the Sheridan VORTAC, the student should note the time, turn to intercept the 302° radial outbound while resetting the bearing selector, and begin the descent to 7,000 feet. When established outbound, the student should begin timing and fly for approximately two minutes before beginning the procedure turn. The procedure turn heading is

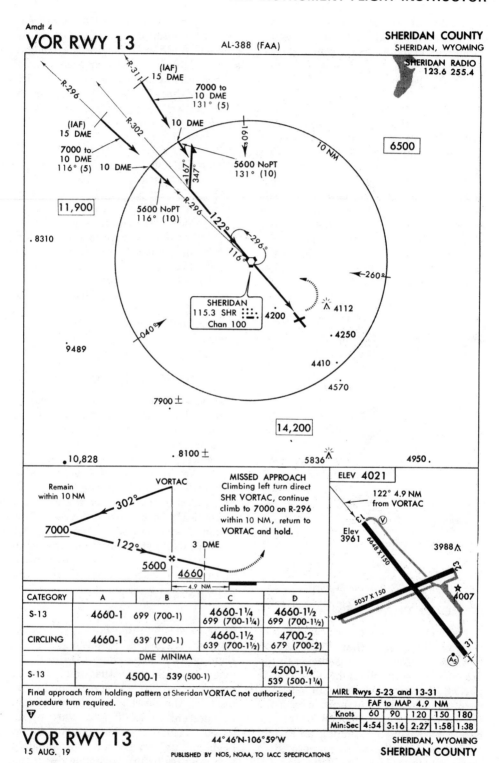

Amdt 4

VOR RWY 13

AL-388 (FAA)

SHERIDAN COUNTY
SHERIDAN, WYOMING

SHERIDAN RADIO
123.6 255.4

(IAF)
15 DME
R-311

7000 to
10 DME
131° (5)

10 DME

R-302

6500

(IAF)
15 DME

7000 to
10 DME
116° (5)

10 DME

R-296

5600 NoPT
131° (10)

10 NM

347°

167°

11,900

5600 NoPT
116° (10)

R-296

122°

-296°

8310

116°

260°

SHERIDAN
115.3 SHR
Chan 100

4200

4112

-040°

4250

9489

4410

4570

7900 ±

14,200

10,828

8100 ±

5836

4950

MISSED APPROACH

Remain
within 10 NM

VORTAC

302°

7000

122°

5600   4660

3 DME

4.9 NM

MISSED APPROACH
Climbing left turn direct
SHR VORTAC, continue
climb to 7000 on R-296
within 10 NM, return to
VORTAC and hold.

ELEV 4021

122° 4.9 NM
from VORTAC

Elev
3961

6648 X 150

3988

5037 X 150

4007

5037 X 150

31

A5

CATEGORY	A	B	C	D
S-13	4660-1	699 (700-1)	4660-1¼ 699 (700-1¼)	4660-1½ 699 (700-1½)
CIRCLING	4660-1	639 (700-1)	4660-1½ 639 (700-1½)	4700-2 679 (700-2)
DME MINIMA				
S-13	4500-1	539 (500-1)		4500-1¼ 539 (500-1¼)

Final approach from holding pattern at Sheridan VORTAC not authorized,
procedure turn required.

MIRL Rwys 5-23 and 13-31

FAF to MAP 4.9 NM

Knots	60	90	120	150	180
Min:Sec	4:54	3:16	2:27	1:58	1:38

VOR RWY 13

15 AUG. 19

44°46'N-106°59'W

PUBLISHED BY NOS, NOAA, TO IACC SPECIFICATIONS

SHERIDAN, WYOMING
SHERIDAN COUNTY

Fig. 4-27. VOR Approach — Sheridan County

maintained for approximately one minute depending on wind before reversing course. During the procedure turn, the instructor should verify that the student has reset the bearing selector to the inbound course and has leveled off at 7,000 feet. The student also should be taught to lead the turn inbound by watching the movement of the CDI.

After the procedure turn, the airplane is established on the inbound radial and a descent to 5,600 feet achieved. The instructor should verify that the student knows the MDA and the missed approach procedure. Crossing the VORTAC inbound, the student should note the time, begin descent to the MDA, and report to Sheridan radio. The instructor should require the student to announce arrival at the MDA, as well as the missed approach point. The instructor may allow the student to land from the approach or require execution of the missed approach procedure. If additional approach training is desired, the latter option usually is chosen. However, it may be beneficial to give the student a brief rest on the ground before resuming the lesson. This also allows the instructor to provide a detailed critique of the student's performance.

The importance of accurate timing and precision flying during the VOR approach must be emphasized to the instrument student. The instructor should point out that, although the VOR approach is termed a nonprecision approach, this does not mean it should not be flown with as much precision as an ILS approach. The instructor should insist the position of the flight be known at all times even when radar vectors are provided by ATC.

The instrument student should understand that the missed approach must be executed immediately upon reaching the MAP if the runway environment is not in sight. Also, the student must maintain the MDA during circling maneuvers until the airplane is in a position from which a normal approach to the runway of intended landing can be made.

During a circling maneuver, the student should understand that the aircraft must remain within the protected area appropriate to its category. The visibility minimum should not be applied as a distance criteria for circling. If at any time during the circling maneuver the student is unable to keep the airport in sight, the missed approach procedure must be executed immediately.

USE OF VOR/DME

The notation "VOR/DME" on an approach chart indicates that DME equipment is mandatory to fly the approach. In many instances, the use of DME equipment and procedures eliminates the need for a procedure turn. DME also provides very accurate position information throughout the approach. Since the approach is based on DME, a time from the FAF to the MAP may not be provided on the chart. In many cases, the MAP is a DME fix, as is the FAF.

The techniques involved in a VOR/DME approach are very similar to those used for a VOR procedure but with the addition of the DME requirements. Should either the ground or airborne DME equipment fail during an approach, a missed approach must be executed at once.

DME ARCS

Since most basic navigation exercises are keyed to following an ADF needle or course deviation indicator, special instruction in flying DME arcs usually is required. To introduce the DME arc, it is best to separate this instruction from an actual approach procedure. In this manner, the student can concentrate solely on flying the arc until proficiency is gained.

Training exercises for DME arcs may be established using a VORTAC, or VOR/DME facility. Initially, an arc can be flown at a moderate distance from the facility and, as the student becomes proficient, the radius of the arc can be reduced. Finally, the student can be assigned a specific radial for termination of the arc, followed by a track inbound to the facility. The FAA recommends the use of a radio magnetic indicator (RMI) for DME arcs unless the pilot is highly proficient in the use of other airborne equipment and in performing the specific approach.

To intercept the arc at groundspeeds of 150 knots or less, the student should begin the turn when one-half nautical mile from the desired DME arc. This technique allows ample distance for the radius of turn and normally prevents undershooting the arc. After the first 90° of turn is completed, the student must realize that a circular path over the ground cannot be flown perfectly. Instead, a series of short, straight segments are used which resemble the arc. This technique is accomplished by continually changing the OBS by 10° in the direction of the arc. In this way, the aircraft is nearly 90° to the selected radial and the orientation to the arc is maintained. Figure 4-28 illustrates the flight path for a typical DME arc.

Wind correction is also a common difficulty during DME arc training. The student should be aware that the wind correction angle never remains constant during the arc. Variations in the DME distance should be used as indications of wind direction and drift. As wind direction becomes obvious to the student, the required correction should not exceed 20° of heading change. If the corrections are too large, difficulties tend to compound themselves rapidly, especially during DME arcs executed close to the facility where the radius is very small. The student should understand that DME arcs are used for transition in many instrument approach procedures; not only VOR/DME procedures, but ILS approaches as well. The main advantage of the DME arc is

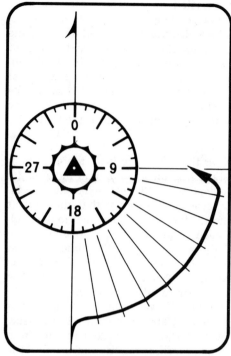

Fig. 4-28. DME Arc Flight Path

that it usually eliminates the need for a procedure turn.

VOR/DME APPROACH PROCEDURE

Figure 4-29 shows the VOR/DME runway 27R approach to Billings Logan International Airport, Billings, Montana (Jeppesen format). The limits of the DME arc are marked by the 040° radial and the 157° radial. In many cases, these radials coincide with airways, so a student approaching on any airway within this span could intercept the 16 DME arc and proceed with the approach.

Assuming the flight is approaching from the east, the turn to intercept the arc should begin at 16.5 DME to avoid overshooting the arc. By continually changing the OBS by 10° increments and turning the airplane 10° in the direction of the arc, the desired flight path can be maintained. The minimum altitude during the DME transition is 5,700 feet until the 078° radial

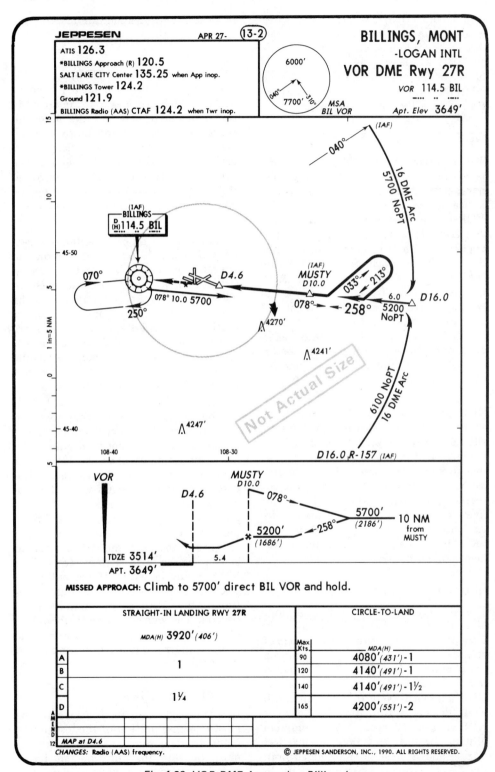

Fig. 4-29. VOR-DME Approach — Billings Logan

(258° inbound course) is intercepted. When inbound, the pilot should descend to 5,200 feet for the intermediate segment. At the 10 DME fix (FAF), the final descent is begun to the appropriate MDA. The 4.6 DME fix is the MAP in this procedure, so timing from the FAF is not required.

NDB APPROACHES

The accuracy of an NDB approach depends on the student's skill using ADF navigation. Although the ADF bearing indicator has some advantages over the VOR indicator, ADF course interception and accurate wind correction are not as apparent. Therefore, an attempt to introduce ADF approach procedures before the student has mastered basic ADF interception and tracking exercises will be frustrating for both the student and the instructor. If considerable time is devoted to basic ADF navigation exercises prior to approach training, better results will be obtained in less time for this important area of instrument flying.

NDB, like VOR approaches, may utilize either on-airport or off-airport facilities. In addition, they may provide for straight-in landings and/or circling maneuvers. Although most large terminal airports utilize NDB approaches, NDBs often are encountered at small airports which are relatively remote from VHF navigation aids. At many of these locations, NDBs provide the only means of executing an instrument approach.

NDB APPROACH PROCEDURE

The NDB runway 9 approach to Kewanee Municipal Airport, Kewanee, Illinois (NOS format), shown in figure 4-30, is typical of on-airport facilities. Assume a training flight is proceeding from Vicks Intersection to Kewanee Municipal Airport for the NDB approach. Note that the approach chart depicts a transition to Kewanee NDB via the 034° bearing. The minimum altitude for this transition is 2,400 feet (MSL), and it *must* be maintained until the flight crosses the Kewanee NDB. In

this case, the transition altitude and procedure turn altitude are the same.

As the flight crosses the NDB, the time is noted and a left turn is made to a heading of 232° which provides a 45° intercept to the outbound course. As the outbound course is intercepted and flown, the student must pay particular attention to wind drift corrections that may be required. The instrument instructor should have stressed the importance of accurately setting the heading indicator before the approach.

Required radio reports should be made only after other items of navigation and aircraft control are completed. The instructor must point out the need to allow sufficient time for establishing the aircraft on the outbound bearing. A hastily flown NDB approach will, more often than not, lead to a missed approach. At the same time, the 10-mile limit for the procedure turn must not be exceeded.

As the student completes the procedure turn, outbound wind corrections should be applied to the inbound heading. The MDA should be checked and the missed approach procedure reviewed. The descent to MDA is begun when the procedure turn is completed and the airplane is established on the inbound bearing. The descent should be rapid enough to ensure reaching the MDA prior to arriving at the MAP and the level-off for MDA should be led by 10 percent of the descent rate. In the procedure illustrated, the MAP is the nondirectional beacon. The missed approach should be started immediately after crossing the NDB if the runway environment is not clearly visible.

ILS APPROACHES

ILS approach training is probably the most exacting portion of the instrument flight training program. The instructor must be sure that the student has a sound background in basic airplane con-

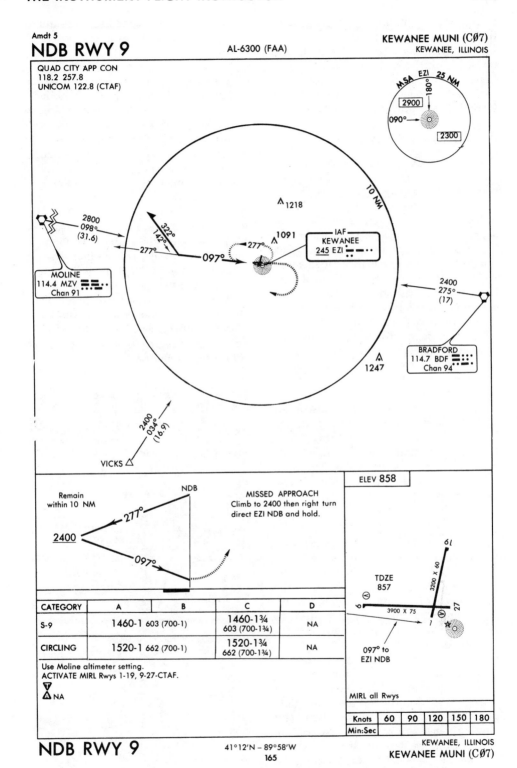

Fig. 4-30. NDB Approach — Kewanee Municipal

trol prior to beginning the training, since the approach requires precise coordination of heading, airspeed, pitch, and power. If difficulties develop in any of these areas, it is recommended the student and instructor digress to basic airwork until the problems are resolved.

ILS APPROACH PROCEDURE

The preferred method of introducing ILS approaches is to introduce each component separately. Specifically, localizer tracking should be learned first, followed by airspeed and rate of descent control on the glide slope. Initially, the instructor should have the student track the localizer, emphasizing precise heading control and the need for small corrections. One method of introducing and developing proficiency in localizer tracking is through the use of holding patterns. The instructor should have the student enter a holding pattern at the outer marker using the localizer as the inbound course. In this way the student acquires proficiency in interception and tracking the localizer. The instructor must emphasize the narrowness of the localizer course which, in turn, makes the CDI four times more sensitive than when tracking a VOR·radial. The need for making very small heading corrections (generally not more than five degrees) should be stressed.

After the student gains proficiency in flying the localizer, descent on the glide slope should be included in the training. The instructor should point out the importance of having the approach airspeed and altitude stabilized prior to glide slope interception, as well as having determined an approximate power setting that will create the desired rate of descent. In this manner, the approach becomes a matter of precise control rather than experimentation. As glide slope interception occurs, pitch and power changes are made to establish the required rate of descent. If a retractable gear aircraft is being flown, lowering the landing gear at glide slope interception

usually creates enough drag to establish the descent with little or no power or pitch change.

Proper aircraft trim technique is especially important during the ILS approach. As the descent stabilizes, power adjustments are used to maintain a constant airspeed and pitch changes are made to remain on the glide slope. These changes should be small and, generally, should be used in conjunction with each other.

Figure 4-31 depicts a training flight inbound on V-19 to Pueblo at 9,000 feet. For the purposes of illustration, the flight is approaching Hanko Intersection when it is handed off to Pueblo Approach Control. After the weather, wind direction and velocity, and altimeter setting are provided, the following clearance is received. *"Cessna 3324R, cleared for the ILS 26 right approach, contact Pueblo Tower 119.1 Aruba inbound."*

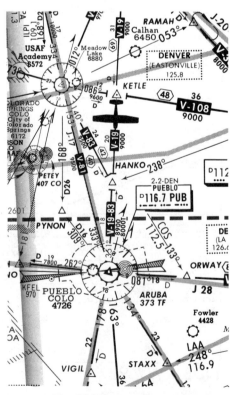

Fig. 4-31. Enroute Segment

Figure 4-32 depicts the ILS runway 26R approach to the Pueblo Memorial Airport, Pueblo, Colorado. Several transition routes are provided, depending on the position of the inbound airplane. These include the 10-mile DME arc and the VOR and NDB transition routes. In this situation, the airplane is located over the Hanko Intersection and the transition will be accomplished via the NDB bearing (161°) to the Aruba outer compass locator (LOM).

After approach clearance is received as the flight passes Hanko Intersection, the student may begin a descent to the published transition altitude of 8,900 feet. This also is an opportunity to slow the aircraft to approach speed and to make other preparations for the approach as early as practicable. The instructor should point out the advantage of developing techniques and habits that will be needed in the future to fly high-performance aircraft.

As the flight crosses Aruba, the student should note the time, plan to fly outbound for two minutes, and turn to intercept the localizer course inbound. The procedure turn is flown normally and, when the student is established inbound, a descent to 6,600 feet is begun. This is the glide slope intercept altitude for this approach procedure. Interception takes place outside the LOM. The instructor should be alert for student problems such as poor scan, not planning ahead, overcontrolling, or confusion. If the approach is made at the end of a long cross-country, fatigue will certainly be an element affecting student performance.

As the flight crosses Aruba inbound, the time should be noted. Although time is not used to determine the MAP on an ILS approach, the approach could be continued as a localizer-only approach with higher minimums if the time to MAP is noted and the glide slope fails. The student should thoroughly understand these options.

When the decision height (which is the MAP on an ILS) is reached, the instructor should call out "runway in sight." If the lesson plan calls for a missed approach, the phrase "no contact" should be used. The student should execute the missed approach at once. During a missed approach, full attention should be devoted to flying the aircraft. The importance of this cannot be overemphasized.

ILS training must include back course approaches, and the student must have a complete understanding of reverse needle sensing. This phase of training may be introduced at any time during ILS approaches to provide variety, dispel monotony, and keep student interest high.

RADAR APPROACHES

Radar approaches can be very useful and, if at all possible, should be a part of every instrument student's training. If a terminal radar facility is not available for this training, the instrument instructor should simulate radar approaches to provide at least a basic familiarity.

There are two types of radar approaches: precision approach radar (PAR) and airport surveillance radar (ASR). PAR approaches provide highly accurate navigation guidance in azimuth and elevation to the pilot. They are available only at military facilities and are available to civil aircraft for emergency use only. ASR approaches are provided by all FAA radar-equipped approach control facilities and, in some instances, by an air route traffic control center.

RADAR APPROACH PROCEDURE

The instructor should explain that the ASR approach is a nonprecision approach. No glide path information is provided during an ASR approach. If requested, however, the controller will provide the recommended altitudes each mile on final, based on the established descent gradient, down to the last mile that is at or above the MDA.

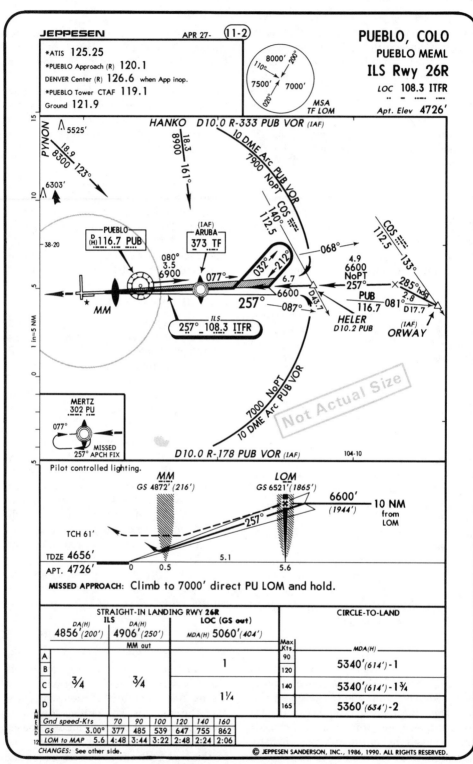

Fig. 4-32. ILS Approach—Pueblo Memorial

The student should be aware that all controller instructions are mandatory. Acceptance of a radar approach does not relieve the pilot of the responsibility for observing the prescribed visibility minimums for the approach. The instrument student also should know that the controller must be advised when visual contact is established with the runway. In addition, the instructor should ensure the student is familiar with "no-gyro" approach procedures.

CONTACT AND VISUAL APPROACHES

Although contact and visual approaches are not published instrument approach procedures, they expedite IFR traffic during the approach phase. Students should be aware that when their destination airport is VFR they may retain their IFR clearance and receive permission to navigate to the runway by visual references on either a contact or visual approach. Both types of approaches relieve the pilot of the requirement to fly the published instrument procedure. Contact and visual approaches can be made only to airports that have standard or special instrument approach procedures.

Contact approaches are initiated at the pilot's request when the destination airport has reported at least one statute mile visibility. The pilot must be able to maintain one mile visibility to the airport and remain clear of clouds. ATC provides separation from other IFR traffic and special VFR traffic. The pilot must maintain separation from VFR traffic as well as obstruction clearance.

For a visual approach, the ceiling and visibility at and in the vicinity of the destination airport must be such that the entire approach and landing can be accomplished in VFR conditions. When ATC radar is employed to separate and sequence air traffic, clearance for a visual approach may be issued only when the aircraft is either at the minimum vectoring altitude or able to descend to that altitude in VFR

conditions. It is the pilot's responsibility to advise ATC as soon as possible if a visual approach is not desired.

Although a pilot may request visual approaches, it is very common during VFR weather conditions for ATC to initiate them in an effort to expedite the flow of traffic to an airport. A visual approach clearance is initiated by the controller when the pilot reports the airport or the preceding aircraft in sight. In the event the pilot has the airport in sight, but does not see the preceding aircraft, the controller can still issue a visual approach clearance. However, in this situation the controller retains the responsibility to maintain aircraft separation and wake turbulence separation. If the pilot reports the preceding aircraft in sight, it then becomes a pilot responsibility to maintain separation from the preceding aircraft and avoid associated wake turbulence. At airports without an operating control tower and no weather reporting facility, ATC may authorize a visual approach only if the pilot advises the controller that the descent and landing at the destination can be completed in VFR conditions.

Additionally, since a clearance for a visual approach may be issued by a remote control facility or by an air route traffic control center, it is important that pilots cancel their IFR flight plans with the appropriate ATC facility. Collision avoidance and obstruction clearance also are the pilot's responsibility during a visual approach.

PRACTICE APPROACHES

When a student has gained reasonable proficiency in holding patterns and instrument approach procedures, two other training procedures of considerable value may be implemented. These are option approaches and instrument approaches begun from a holding pattern. In both cases, realistic situations can be simulated with considerable accuracy.

OPTION APPROACH

The option approach is available only at airports with an operating control tower. If ATC approves the option approach, the instructor is provided with several choices. The student may be required to make a full stop landing, a touch and go, a rejected landing, or the missed approach as published or as instructed by ATC. This provides an excellent opportunity to observe the student's planning ability and reactions to unexpected situations.

APPROACHES FROM HOLDING PATTERNS

Having the student hold at the final approach fix and begin the approach from the holding pattern prepares the student for typical ATC procedures used during actual instrument flight. When this occurs in training, regardless of the position in the holding pattern, the student is expected to cross the final approach fix and proceed inbound without executing a procedure turn. The only exception to this procedure is when the published approach procedure states that final approach from a holding pattern is not authorized. In this case, the student must comply with the published procedure.

IFR CROSS-COUNTRY

IFR cross-country work can begin after the student develops proficiency in approaches and in receiving, understanding, and executing ATC clearances. It is during this portion of the training that the student should be able to combine all the previously learned maneuvers and procedures into a meaningful cross-country flight under instrument flight rules. It is important that the instructor ensure the student is competent and ready for the enroute phase of training. If the student is weak in any of the previous phases of training, the cross-country flight may be meaningless to the student and frustrating for the instructor.

PLANNING CONSIDERATIONS

The IFR cross-country flight requires thorough and accurate planning for all phases of the flight. In addition to the normal planning of checkpoints, the instructor should emphasize the following areas not only as a matter of accurate planning, but also as a matter of judgment.

1. SID acceptance
2. Route selection
3. Alternate airport and fuel requirements

SID ACCEPTANCE

Prior to the cross-country flight, the student should be fully aware of the advantages and disadvantages of the standard instrument departure. The purpose of any SID is to standardize the departure path and eliminate radio frequency congestion. The instructor must emphasize that certain departures cannot be accepted due to the performance limitations of the airplane. For example, figure 4-33 shows the relatively simple Maric Three Departure with the Gorman Transition. If the flight departs on runway 30L or 30R, the airplane simply turns left to a heading of 240°, intercepts the Shafter 194° radial, proceeds to the Maric Intersection, and then on course. The instructor must point out the notation section of the procedure which states that the transition requires a minimum climb rate of 220 feet per nautical mile to 9,500 MSL. Many students misconstrue this notation to mean 220 feet per minute. Therefore, if the airplane has a climb speed of 100 knots, the required climb rate would be 367 feet per minute to 9,500 MSL. The student should understand the requirements in terms of airplane performance capabilities of any SID prior to acceptance.

ROUTE SELECTION

Although the selection of an IFR route is a fundamental decision, certain aspects are worthy of notation. First, due to the altitude requirements of many MEAs

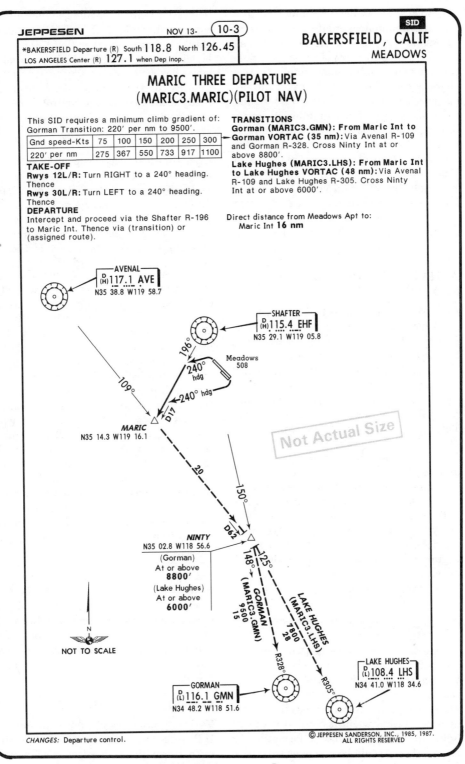

Fig. 4-33. Standard Instrument Departure

and MOCAs, it is not uncommon to find that the minimum enroute altitude is in excess, or very close to the service ceiling of the training airplane. In this situation, the student should understand that a slightly greater distance between the departure and arrival airport is better than the additional fuel consumption during a prolonged climb. Therefore, the shortest distance between two points may not be the most practical situation in terms of time enroute and overall fuel consumption. Although this situation prevails in only a few areas, the instructor should include a short discussion on airplane performance versus enroute altitude limitations.

ALTERNATE AIRPORT AND FUEL REQUIREMENTS

After the route is selected, the student must determine if an alternate airport is required. The instructor should carefully review the weather requirements of FAR Part 91.169 with the student and make the determination. In the event an alternate is required, the student must select an airport within the fuel range of the airplane. This aspect becomes particularly important during long cross-country flights where it may be possible to fly to the destination and to the alternate but not have the 45-minute reserve required by regulations. The student should apply this regulation to the practical aspect that fuel for holding delays and approaches must be accounted for accordingly. It is not uncommon for beginning students to choose alternate airports that are outside the range parameters of the airplane.

IFR CLEARANCES

An instrument instructor has the responsibility of teaching students the most expeditious means of receiving a clearance. First, the student should realize that the clearance very often is issued exactly as requested. If the student is familiar with the order in which clearance information is issued, there should be no major surprises from ATC.

Special attention should be devoted to copying the initial clearance. Pretakeoff checks, taxiing, or other flight duties can be distracting and should be minimized while the clearance is being copied. If ATC informs the student that the clearance is ready while taxiing, the controller may be requested to "stand by" until the aircraft is stopped and the student is ready to copy.

After the clearance is copied, it should be reviewed for any unexpected changes, especially in respect to clearance limits, intersections, or checkpoints which are unfamiliar. Also, the student should ensure that the assigned communications frequencies are within the capability of the aircraft radios.

The student should ensure the initial clearance has been copied correctly by reading it back to the controller verbatim. When airborne, the student must read back at least those parts containing altitude assignments or vectors as a means of mutual verification. Numbers should be read back in the same sequence they were received, and students should always include the aircraft identification when acknowledging or reading back a clearance. In some cases, the controller may specifically request that the clearance be read back if it is unusually complex, radio reception is poor, or confirmation is necessary for any other reason. A clearance should be read back when the student has any element of doubt concerning the clearance.

After the initial clearance has been received, it may be necessary for ATC to modify or add to the clearance, or the student may request an amendment. In either case, the various conditions and instructions in the last clearance received have precedence over any related items in previous clearances.

ENROUTE PROCEDURES

Various procedures should be included in IFR cross-country training, including holding patterns at a VOR-established intersection using only one navigation receiver both full- and partial-panel flight, dead reckoning navigation, and simulated in-flight emergencies. FAR 61.65 requires one instrument cross-country on Federal airways or as routed by ATC, that is at least 250 nautical miles including VOR, ADF, and ILS approaches at different airports.

The instructor should require the student to know the position of the flight at all times. The instructor can develop good navigation habits in each student by occasionally asking for an estimate to the next fix or the flight's current position relative to the last fix. The student must be required to continue careful and accurate navigation even while in radar contact. As cross-country training progresses, communication failures should be introduced. Normally, this can be simulated during a flight to an uncontrolled airport. Prior to their introduction, the requirements of FAR Part 91.185 should be discussed in detail.

REPORTING REQUIREMENTS

Although radar negates the need for standard position reporting, certain reports should be made while the airplane is in radar contact. Required reports, whether in a radar or nonradar environment, are summarized in figure 4-34.

Additional reports are required whenever any of the following equipment malfunctions occur.

1. Loss of VOR, TACAN, or ADF receiver capability
2. Complete or partial loss of ILS receiver capability
3. Impairment of air/ground communications capability

The pilot should advise the controller of the degree to which the flight's operational capabilities are limited. Also, the type and extent of assistance that will be required from ATC should be specified.

Nonradar Procedures

Before the installation of radar facilities, aircraft separation was based on different altitude assignments, different routes, and time/distance intervals between flights. Although less effective in terms of handling large volumes of air traffic, this system is still used today on some low altitude routes or in the event of a radar outage.

Since radar and transponder signals operate on the line-of-sight principle, signal blockage and low altitude dead spots may create areas where radar coverage is not adequate for safe separation. Also, airborne transponder failure or severe weather conditions may degrade radar reception and prevent positive aircraft identification.

Position Reports

To maintain separation in the nonradar environment, ATC must be provided with frequent position and altitude reports. In addition, estimated times of arrival over succeeding navigation aids or fixes must be accurately calculated and reported to ATC by the pilot. Therefore, navigation logs should be updated continuously. This is equally important for operations conducted in radar contact, since radar service may be terminated.

FAR Part 91.183 requires pilots to maintain a "listening watch" on appropriate frequencies during all IFR operations. In addition, pilots must furnish position reports to ATC without request when crossing compulsory reporting points when not in radar contact. An example of the standard position report format used for nonradar operations is shown in the center of figure 4-34. On initial callups, the pilot should specify the facility call sign, aircraft identification, and location. The pilot should then wait for a reply *before* giving the report.

CENTER COMMUNICATIONS

In most instances, the distribution of remote communications outlets (RCOs) is adequate for direct pilot/controller communications. Each sector controller or team has its own discrete frequency. These frequencies are shown on IFR enroute

COMPULSORY IFR REPORTS — RADAR/NONRADAR

Vacating one assigned altitude or flight level for another	*"Saratoga 6758L, leaving 7,000, climbing to 10,000."*
VFR on top change in altitude	*"Saratoga 6758L, VFR on top, climbing to 10,500."*
Unable to climb/descend at 500 feet per minute	*"Saratoga 6758L, maximum climb rate 400 feet per minute."*
Missed approach	*"Saratoga 6758L, missed approach, request clearance to Omaha."*
TAS variation from filed of 5%, or 10 knots, whichever is greater	*"Saratoga 6758L, advises TAS decrease to 150 knots."*
Time and altitude or flight level reaching a holding fix or point to which cleared	*"Saratoga 6758L, Fargo Intersection at 05, 10,000, holding east."*
Leaving any assigned holding fix or point	*"Saratoga 6758L, leaving Fargo Intersection."*
Loss of nav/com capability	*"Saratoga 6758L, ILS receiver inoperative."*
Unforecast weather conditions or any other information relating to safety of flight	*"Saratoga 6758L, experiencing moderate turbulence at 10,000."*

STANDARD POSITION REPORT — NONRADAR

Identification	*"Saratoga 6758L,*
Position	*Shreveport,*
Time	*15,*
Altitude/flight level	*11,000,*
IFR or VFR for report to FSS only	*IFR,*
ETA over the next reporting fix	*Quitman 40,*
Succeeding reporting points	*Scurry next."*
Pertinent remarks	*(Infrequently used)*

ADDITIONAL REPORTS — NONRADAR

Leaving FAF or OM inbound on final approach	*"Saratoga 6758L, outer marker inbound, leaving 2,000."*
Revised ETA of more than three minutes	*"Saratoga 6758L, revising Scurry estimate to 55."*

Fig. 4-34. IFR Reports—Radar Environment

charts. However, flight plan filing and requests for weather or similar data should be made through the nearest FSS. In cases where direct pilot/controller communications are lost and cannot be reestablished, the student should be instructed to furnish the IFR position reports to the closest FSS. The FSS specialist then provides a communications relay between the student and the ARTCC controller.

ADVANCED INSTRUMENT TRAINING

After a student has received an instrument rating, the problem of maintaining proficiency must be considered. If the pilot regularly flies on a flight plan in instrument conditions, proficiency can be maintained reasonably well. However, this is not the case with many pilots. For the majority of pilots, the instrument competency check is the best way to ensure currency. If a pilot needs additional proficiency training, instrument refresher instruction is the answer.

A refresher flight should include a thorough oral discussion of regulations and procedures to ensure currency of knowledge. The flight portion should involve a review of departure, enroute, holding, and approach procedures, including missed approaches. Full- and partial-panel flight in all regimes, including critically slow airspeeds, stalls, and recovery from unusual attitudes also should be included.

INSTRUMENT COMPETENCY CHECK

The instrument competency flight check is an alternate method for the instrument rated pilot to *regain* instrument currency. This competency check may be used to obtain instrument currency in lieu of six hours of instrument flight in the past six months, but it becomes *mandatory* if the pilot has not been current at any time during the preceding 12 months.

The instrument competency check, like the flight review, should be a learning experience, as well as an evaluation session for the pilot. The pilot should be instructed on new instrument procedures, techniques, and regulation changes. In addition, a thorough review of previously learned skills should be conducted.

The airplane used for the check should meet the IFR equipment requirements of FAR Part 91 and should have adequate avionics to perform both precision and nonprecision instrument approaches. As an alternative, all or part of the check may be performed in a ground trainer or simulator which is authorized by the FAA.

Prior to the competency check, the pilot should review appropriate subject areas in preparation for the oral discussion. Audiovisual materials provide an excellent method of review in a minimum of time. Additional review may include selected chapters from a current instrument rating or instrument/commercial textbook, and appropriate FAA publications.

In addition, the instructor should provide a series of essay-type questions for the pilot to research prior to the scheduled oral discussion. However, the preflight oral discussion should not be limited to the questions provided. The pilot or instructor should expand the scope of a given question or introduce new subject areas, wherever appropriate.

The instrument competency check normally begins with the oral briefing and discussion session. As a minimum, the pilot should be quizzed on ATC procedures, FAR Part 91, IFR flight planning, and aircraft performance. If the pilot's general knowledge in a given area appears to be weak, it should be discussed and reviewed in sufficient detail to ensure understanding. In addition, the pilot may be encouraged to review specific audiovisual subjects.

Following the oral discussion, the pilot should prepare and file an IFR flight plan to a point outside the local area. The planned flight should take into account the equipment and performance capabilities of the airplane.

The actual flight check should consist of selected procedures and maneuvers from the instrument rating practical test standards to evaluate the pilot's performance in each of the required areas of instrument competency. As in the ground session, if the pilot does not display the required proficiency in any area, it often is beneficial to explain, demonstrate, and review the procedure or maneuver. This way, skills can be increased to the necessary levels, and the need for a second check frequently can be avoided.

Successful completion of the check provides immediate instrument currency for the pilot, and an appropriate logbook entry should be made. If the competency check is not satisfactory, the logbook entry should indicate a dual instruction period. At no time should an entry indicate an unsatisfactory instrument competency check.

chapter 5

THE MULTI-ENGINE FLIGHT INSTRUCTOR

INTRODUCTION

This chapter addresses those individuals who desire the addition of a multi-engine class rating to their instructor certificates. Consequently, it is assumed the applicant is at least a commercial pilot with instrument and multi-engine ratings and also possesses a basic flight instructor certificate with airplane, single-engine land privileges. Some of the material presented is considered to be a review of multi-engine knowledge and procedures, while the rest is oriented toward the needs of the beginning multi-engine instructor. The first section presents a detailed analyses of multi-engine operations and engine-out teaching procedures and concludes with important safety considerations for the instructor. The second section covers the entire range of multi-engine and engine-out aerodynamics. This material also is appropriate for the beginning instructor since many operational procedures stem from aerodynamic considerations. The entire chapter is based on training in multi-engine airplanes with wing-mounted engines, either conventional or counter-rotating.

SECTION A—MULTI-ENGINE AIRPLANE INSTRUCTION

As the instructor begins work with a multi-engine student, it is beneficial to organize the training program into two basic phases. First, the student must transition into the multi-engine training airplane and master all the basic flight maneuvers. Second, after the student has demonstrated the safe and accurate maneuvering of the airplane with both engines operating, the instructor may introduce and teach engine-out and emergency procedures. This section adheres to this philosophy and guides the instructor applicant through the teaching process as it is applied to multi-engine training. The order in which procedures and maneuvers are presented in this section does not necessarily follow the order used during training. The training sequence is determined by the instructor and the training syllabus.

AIRPLANE TRANSITION

The instructor's basic goal during the transition from the single-engine to the multi-engine airplane, is to develop the student's perceptions. Generally, the student is transitioning to an airplane that has "heavier" control responses, unfamiliar instrumentation and outside visual references, a higher speed range, and, in most cases, more complex systems.

The first step in smooth airplane transition is a thorough ground training session. This briefing should include a review of the appropriate pilot's operating handbook, with emphasis on the sections that are important for the first flight lesson. It is appropriate to assign such sections as engine starting, taxiing, and "V" speeds as prerequisites to the first flight lesson. In addition, the student should be familiar with take-off planning in terms of single-engine service ceiling and accelerate-stop distance computations.

The second step is to familiarize the student with the exterior and interior components of the airplane. The exterior preflight inspection should be conducted by referring to the printed checklist provided in the pilot's operating handbook. The interior inspection should include a complete explanation of each switch, dial, gauge, and avionics component. This procedure enables the student to feel more comfortable during the first flight because each item is easily located without unnecessarily diverting attention from the maneuver or procedure being performed. During cockpit familiarization, the student should perform the engine start and engine shutdown checklists.

INTRODUCING THE AIRPLANE

Once the student has had the opportunity to review the concepts of the airplane orientation session and learn the assigned sections of the pilot's operating handbook, the instructor should plan the first flight lesson. This lesson should be devoted to basic airplane familiarization.

TAXI PROCEDURES

After the engines have been started in accordance with the printed checklist and manufacturer's recommendations, correct taxi procedures are introduced. During initial taxi operations, the student should practice using equal power on both engines and rely on the steerable nosewheel for directional control. This aspect is important from the first power application through the entire taxi phase. Typically, the student will attempt to maintain equal power by throttle position. As a correcting measure, the instructor should encourage the use of the engine tachometers as references, and should emphasize that the use of asymmetrical power is a taxi technique to be used only during crosswind conditions or as an aid in reducing the turning radius. However, the student should understand the disadvantages of using differential power include tire wear and side loading on the landing gear. The least desirable method of directional control is differential braking. The student should understand that this technique also increases side loads, wear on both the tires and brakes, and can easily result in brake overheating and subsequent failure.

TAKEOFFS AND CLIMBS

As the student begins to work with the different takeoff techniques and departure climbs, the instructor should emphasize the operational differences between single-engine and multi-engine airplanes. Specifically, the student must understand the significance of the V-speeds required for takeoff and the specific airspeed parameters in which the airplane must be operated.

NORMAL TAKEOFF AND CLIMB

The most common student difficulties during takeoffs stem from uneven power application. Directional control problems also may occur if one engine falters due to rapid power application. Excessive speed during the ground run and the departure climb also are common student tendencies. Often, this is caused by too much attention to engine instruments, improper trim control, or an inaccurate assessment of acceleration. Although these problems are not serious initially, the instructor should take immediate corrective action and remain especially alert during the multi-engine takeoff phase.

Throughout the entire takeoff and climb, the instructor must emphasize that the combined maneuver is an orderly progression of events that are dependent upon precise airspeed control. Single-engine airplane pilots frequently do not place sufficient importance on accurate airspeed control immediately following liftoff; however, multi-engine airplane pilots *must* thoroughly understand that the proper airspeed is critical, particularly in the event one engine loses power.

Therefore, the instructor must stress the various V-speeds applicable to the training airplane. For a normal takeoff, V_{MC} (minimum control airspeed), V_R (rotation speed), V_{LOF} (liftoff speed), V_Y (best rate-of-climb speed), and V_{YSE} (best rate-of-climb speed with one engine inoperative) are the most important considerations. In this manual, the FAA term V_{MC} is used for minimum control airspeed instead of V_{MCA} which is preferred by some manufacturers. When V_{MCA} is used, V_{MCG} also may be used to indicate ground minimum control speed. Another V-speed established for some light twins is V_{SSE}. This speed is defined as the minimum speed selected by the manufacturer for intentionally rendering one engine inoperative in flight for pilot training. In all V-speeds, the "V" represents velocity and those ending with "SE" refer to engine-out operations. V-speeds are defined and discussed further throughout this chapter.

NOTE:

V_{SSE} is predicated upon the maintenance of conservative controllability margins when one engine is suddenly and intentionally rendered inoperative. Its selection is based upon the characteristics of the specific airplane to which it applies; however, in no case may it be lower than $1.05\ V_{MC}$.

The airspeed for rotation on takeoff must be appropriate to the individual airplane. The FAA recommends liftoff at $V_{MC} + 5$ knots, or the manufacturer's recommended liftoff airspeed. If the manufacturer's recommendation is not available, the airplane should not leave the ground on a normal takeoff prior to V_{MC} and, preferably, $V_{MC} + 5$ knots, as shown in figure 5-1. The landing gear should be retracted

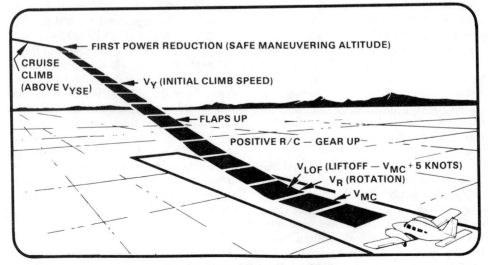

FIRST POWER REDUCTION (SAFE MANEUVERING ALTITUDE)

CRUISE CLIMB (ABOVE V_{YSE})

V_Y (INITIAL CLIMB SPEED)

FLAPS UP

POSITIVE R/C — GEAR UP

V_{LOF} (LIFTOFF — $V_{MC} + 5$ KNOTS)

V_R (ROTATION)

V_{MC}

Fig. 5-1. Normal Takeoff and Climb

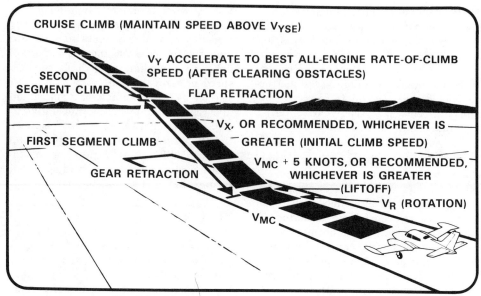

Fig. 5-2.   Maximum Performance Takeoff and Climb

after a positive rate-of-climb has been established and when it is no longer practical to land on the remaining runway. After the flaps are raised, the pitch attitude should allow the airplane to accelerate to the best all-engine, rate-of-climb speed. V_Y should then be maintained with takeoff power until a safe maneuvering altitude is attained.

MAXIMUM PERFORMANCE TAKEOFF AND CLIMB

Maximum performance takeoff and climb training should be introduced after the student has gained a reasonably high level of proficiency in airspeed control and is familiar with the handling characteristics of the airplane. The student also should have a thorough knowledge of climb airspeeds, including the significance of the engine-out speeds appropriate to the training airplane.

The pilot's operating handbook should be used to determine the recommended technique for executing the takeoff and subsequent maximum performance climb. However, some manufacturers recommend rotation and liftoff at a speed *less* than V_{MC}. The instructor should emphasize the hazards and cover the FAA recommended

procedure that the liftoff occur at V_{MC} + 5, or the recommended airspeed, whichever is greater. The student should fully understand the advantages of accelerating to an adequate airspeed on the ground and the consequences of a power loss following early rotation at an airspeed below V_{MC}.

The takeoff should be planned so that liftoff occurs at the correct speed. Liftoff should be followed by a maximum performance climb at V_X, or the recommended speed (whichever is greater) until the obstacle is cleared or to at least 50 feet AGL. Procedures for raising the gear and flaps and accelerating to V_X should follow the pilot's operating handbook recommendations. After that, the airplane should be accelerated to $V_Y \pm 5$ knots to avoid high pitch angles. The climb should be continued until a safe maneuvering altitude is reached. Then, the power can be reduced and the recommended climb airspeed established for the remainder of the climb to altitude. This technique is depicted in figure 5-2.

During the first attempts, the student typically will accelerate past the recommended liftoff speed and maintain an excessive airspeed during the climb. The instructor should teach the student to begin elevator back pressure approximately five knots prior to the desired liftoff speed. In this manner, the airplane rotation is initiated early enough so that liftoff at the correct airspeed is less difficult. As a maximum performance climb is started, many students do not establish an adequate pitch attitude to maintain the correct airspeed. Airspeed control problems are common but easily corrected through instructor preflight and postflight briefings and student practice.

Improper power control is apparent during the power reduction after the obstacle is cleared. As a general rule, the instructor should teach that power should not be reduced for climb before the landing gear and flaps are fully retracted. This technique follows the philosophy that maximum power should be available until the drag is eliminated in the event one engine loses power. Specifically, the instructor should stress that the landing gear is retracted as the airplane achieves a positive rate-of-climb. After the obstacle is cleared and the pitch attitude smoothly reduced to allow acceleration to Vy, the flaps should be retracted. After Vy is established with the gear and flaps fully retracted, the power may be reduced to the climb power setting. The following list includes common student errors that apply to both normal and maximum performance takeoffs and climbs.

1. Improper use of takeoff and climb performance data
2. Improper initial positioning of flight controls or wing flaps
3. Improper power application
4. Inappropriate removal of the hand from the throttles
5. Improper directional control
6. Improper use of aileron
7. Rotation at improper airspeed
8. Failure to establish and maintain proper climb configuration and airspeeds
9. Drift during climb

Maximum performance takeoffs are subject to some other distinctive errors. Students commonly fail to position the airplane for maximum utilization of the available runway area. In addition, they frequently use the brakes improperly.

NORMAL APPROACHES AND LANDINGS

Although normal approaches and landings do not pose significant problems for the multi-engine student, the instructor may find it beneficial to introduce the transition to approach and landing for the first time at altitude. By using this procedure, the student learns the flight characteristics of the airplane in the approach configuration and, in addition, becomes familiar with the power settings required to maintain the desired airspeed and rate of descent.

LANDING APPROACHES   95 mph

The instructor must emphasize the importance of establishing a stabilized power approach when the student begins concentrated work with landings. The technique is most easily introduced at altitude where prolonged descents at different power settings and configurations can be established. Generally, the student should slow the airplane to the approach airspeed, lower the first increment of flaps, and note the power setting required to maintain a constant altitude. Then, the landing gear should be extended while maintaining a constant airspeed. The drag created by the landing gear normally will result in a rate of descent of approximately 300 to 500 feet per minute. After the descent is stabilized at the approach airspeed, the remaining flaps can be extended to establish the final approach configuration. The student should understand that, during a stabilized power approach, the power remains nearly constant and the drag created by the landing gear and flap extension produces the necessary rate of descent. Once the approach is established, *minor* power adjustments may be made, as necessary, to maintain the approach profile. The student should realize the disadvantages of a multi-engine, power-off approach in the landing configuration which are due to the very steep approach attitude and extremely high rate of descent.

Following the exercise in stabilized power approaches, the student must thoroughly understand that the landing approach actually begins prior to entering the traffic pattern. More specifically, the student should have a firm "plan of action" which includes the following techniques.

1. Establishing the required airspeed prior to entering the traffic pattern

2. Knowing the position within the traffic pattern where the landing gear and flaps should be extended

3. Determining the approximate altitude the airplane should achieve at each point within the traffic pattern

Once these three techniques are accurately preplanned, any type of approach to a landing becomes a matter of student prac-tice and instructor followthrough. The entire procedure is illustrated in figure 5-3.

The majority of student difficulties are the product of not preplanning the traffic pattern. This often makes it difficult for the student to slow the airplane to the approach airspeed. For example, if the student inadvertently uses too high an airspeed in the pattern, the result will be large and erratic power reductions and additions. The excessive airspeed will prevent the flaps and gear from being extended on schedule, and the final approach will be too high; thus creating a steep approach angle, excessive rate of descent, and a high probability of a go-around.

The following list includes common student errors that apply to both normal and maximum performance approaches and landings.

1. Improper use of landing performance data and limitations

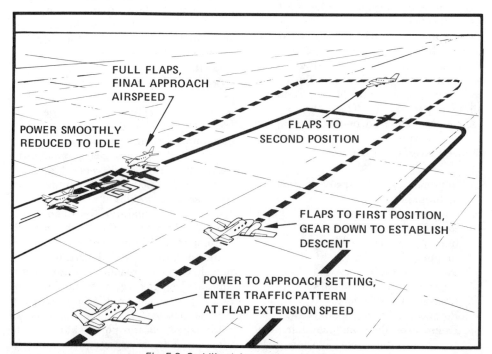

Fig. 5-3. Stabilized Approach to Landing

2. Failure to establish approach and landing configuration at the proper time or in proper sequence

3. Failure to establish and maintain a stablized approach

4. Failure to use the proper technique for wind shear or turbulence

5. Inappropriate removal of the hand from the throttles

6. Faulty technique during roundout and touchdown

7. Poor directional control after touchdown

8. Improper use of brakes

During maximum performance approaches and landings, students commonly display faulty technique in the use of power, wing flaps, and trim. Although these problems are typical, they are corrected easily through conscientious preflight and postflight briefings and student practice.

GO-AROUNDS

The procedures associated with go-arounds are most effectively introduced at altitude. This allows students to become familiar with the correct drag cleanup sequence and learn to maintain altitude and airspeed while their attention is diverted. Go-arounds can be introduced during actual approaches to landings after the student is familiar with the procedure.

To begin the maneuver, the pilot's operating handbook should be consulted to determine the specific procedures and limitations for the training airplane. However, most procedures require that maximum power be applied, followed by flap retraction to the takeoff setting. After a positive rate-of-climb is established, the landing gear and the remaining flaps are retracted. If climb performance is not critical, it frequently is advisable to leave the landing gear extended and retract the remaining flaps after a positive rate-of-climb is established. During the entire maneuver, airspeed should be closely monitored to ensure acceleration to the best rate-of-climb airspeed as rapidly as practical.

These are a number of common student errors which apply to go-arounds.

1. Failure to recognize a situation where a go-around is necessary

2. Hazards of delaying a decision to go around

3. Improper power application

4. Failure to control pitch attitude

5. Failure to compensate for torque effect

6. Improper trim technique

7. Failure to maintain recommended airspeeds

8. Improper wing flaps or landing gear retraction procedure

9. Failure to maintain proper track during climbout

10. Failure to remain well clear of obstructions and other traffic

MANEUVERING DURING SLOW FLIGHT

For the practical test, flight instructor applicants must be prepared to explain slow flight in terms of the relationship of the maneuver to critical flight situations, such as go-arounds. The airplane's flight characteristics and controllability during slow flight are directly affected by configuration, weight, CG, maneuvering loads, angle of bank, and power.

Maneuvering during slow flight provides the student with an opportunity to learn airspeed control in many different configurations, the handling characteristics of the airplane, and the effect of the flaps and landing gear on performance. The student should learn to perform the maneuver at an airspeed of five knots above stall speed or VMC, whichever is greater. The airspeed tolerance is -5 knots. The student should be able to execute the maneuver in straight-and-level flight and during level turns. In addition, the maneuver should be performed with various flap settings in both cruising and landing configurations.

To introduce the maneuver, the instructor should have the student slow the airplane to the appropriate airspeed with the flaps and landing gear retracted. This condition is most easily established by having the student smoothly reduce power and apply adequate elevator or stabilator back pressure to maintain a zero rate of climb or descent. As the airspeed decreases to the desired airspeed, enough power should be added to stabilize the airspeed and altitude of the airplane. After the airplane is stabilized and trimmed for straight-and-level flight, medium banked turns should be made in both directions.

Prior to introducing the effect of flaps and landing gear, the instructor should emphasize how configuration changes affect airspeed and total drag. For example, as the flaps are extended, *induced drag* increases and, when the landing gear is extended, *parasite drag* increases. Therefore, as total drag is increased by flap or landing gear extension, a power adjustment is required to maintain a constant airspeed and altitude. Although this concept is fundamental, it is best demonstrated by first slowing the airplane to the appropriate slow flight airspeed with the flaps and landing gear retracted.

Then, the student should extend the first increment of flaps and adjust the pitch and power, as necessary, to maintain altitude and airspeed. After the airplane is stabilized and retrimmed, the second increment of flaps should be extended, followed by the landing gear and full flaps. Before each flap extension, the student should stabilize the airplane and note the resultant increase in power that is required to maintain altitude and airspeed. To complete the exercise, the student should reverse the process and make the necessary power reductions until the airplane is established in slow flight with the flaps and landing gear retracted. As proficiency improves, the instructor should require the student to perform the entire exercise with minimal variation in altitude and designated airspeeds.

Some of the common student errors during slow flight are shown in the following list.

1. Failure to establish specified configuration
2. Improper entry technique
3. Failure to establish and maintain the appropriate reduced airspeed
4. Excessive variations of altitude and heading when a constant altitude and heading are specified
5. Rough or uncoordinated control technique
6. Improper correction for torque effect
7. Improper trim technique
8. Unintentional stalls
9. Inappropriate removal of the hand from the throttles

Predominant student difficulties during the maneuver include improper attitude and power control resulting in variations in airspeed and altitude. Since larger and heavier airplanes respond slower while maneuvering during slow flight, this maneuver should be performed at altitudes sufficiently high to allow recoveries to be completed at least 3,000 feet above the ground.

STALLS

Instruction in stalls is included in all multi-engine flight training programs for two basic reasons. First, students learn the proper recognition and recovery techniques; and second, they become familiar with the handling characteristics of the airplane as the stall approaches. The pilot's operating handbook recommendations must be followed for possible limitations concerning stalls in the training airplane.

Full stalls using high power settings have been deleted from multi-engine practical tests because the excessively high pitch angles necessary to induce these stalls may result in uncontrollable flight. This is reflected in the multi-engine sections of the private, commercial, and flight instructor PTS. Examiners and instructors

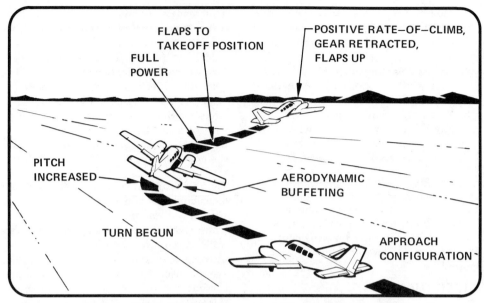

Fig. 5-4. Stall — Gear Down and Approach Flaps Configuration

also are cautioned against performing any stalls with one engine at reduced power or inoperative and the other engine developing effective power. They should also be alert to the possible development of high sink rates when performing stalls in multi-engine airplanes with high wing loadings. Therefore, the loss of altitude during stall entries should be no more than 50 feet. The entry altitude should allow recoveries to be completed no lower than 3,000 feet AGL.

Power-on and power-off stalls in multi-engine airplanes may be performed in the following configurations:

1. Gear up and flaps up

2. Gear down and approach flaps

3. Gear down and full flaps

Recoveries should be initiated at the first indication of buffeting or decay of control effectiveness using the correct power and control application. For the gear down and approach flaps configuration, the instructor may require the student to maintain a ground track appropriate to a simulated traffic pattern. The student first slows the airplane to the V_{YSE},

and then extends the flaps and landing gear as though the airplane were in an actual traffic pattern. As shown in figure 5-4, when the airplane is on the simulated final approach, the student should establish a medium bank, coordinated turn and begin steady elevator back pressure to maintain a zero rate of climb or descent. At the first indication of the stall, the student should apply *slight* forward pressure, smoothly apply full power to both engines, and retract the flaps to the takeoff setting or as recommended by the manufacturer. Once a positive rate-of-climb is established, the landing gear should be retracted and the climb continued to the original altitude. The student may be asked to recover to a specified airspeed and configuration.

Stall recognition and recovery normally does not pose significant problems for the multi-engine applicant because the procedures are basically the same as those used in a high-performance, single-engine airplane. Common student errors regarding stalls are shown in the following list.

1. Failure to establish proper configuration prior to entry

2. Improper pitch or bank attitude during entry
3. Rough, uncoordinated control technique
4. Failure to achieve a stall
5. Poor stall recognition and delayed recovery
6. Poor torque correction
7. Excessive altitude variation during entry or recovery
8. Secondary stall during recovery

Occasionally, the student may not apply power evenly, or one engine may falter momentarily causing severe asymmetrical thrust. The instructor must be very alert for this possibility and take the controls, if necessary, to prevent sudden loss of control.

STEEP TURNS

The steep turn is an excellent maneuver to help the student develop a feel for the control response of the airplane. Because of the high load factors involved, these turns should be performed at a speed which does not exceed the manufacturer's recommended maneuvering speed (V_A). The maneuver should be started from a stabilized airspeed, altitude, and usually a cardinal heading and entry should be smooth and coordinated. The Practical Test Standards specify a 45° (±5°) bank turn for private pilot applicants and at least a 50° (±5°) bank for commercial training. Normally, turns should be continued for 360° and the roll-out should be timed so the wings are level when the airplane returns to the entry heading. Lead for the roll-out should be approximately one-half the bank angle. Roll-in and roll-out should be at the same rate.

The instructor should encourage the use of *moderate* elevator trim to alleviate most of the required back pressure during the steep turn while the student notes the elevator back pressure requirements as the bank increases for entry and decreases for recovery. In addition, the student should note the loss of airspeed as the bank increases and apply adequate power to maintain a constant airspeed, followed by a power reduction as the bank angle decreases during recovery. After the student has reasonable success with 360° turns, the instructor can introduce variations to the maneuver by requiring the student to roll immediately from a steep turn in one direction to a steep turn in the opposite direction. As proficiency is achieved, the student may be required to make continuous alternating steep turns with no more than 180° of change in heading while omitting any hesitation between the opposing turns.

The steep turn is a *visual* maneuver and occasional instrument reference is made *only* for confirmation. However, since many training airplanes do not have good forward visual references for adjusting pitch attitude with the horizon, the instructor should encourage the student to smoothly roll into the turn and occasionally monitor the vertical speed indicator to reference a zero rate of climb or descent. This procedure will help the student overcome the major difficulty of maintaining proper altitude control.

When altitude control becomes a pronounced student error, the instructor should evaluate the student's use of elevator trim. If the student attempts to trim the airplane for a "hands off" 50° banked level turn, the forward pressure during roll-out becomes excessive, and control of the airplane becomes very difficult. Conversely, if elevator trim is completely omitted during the turn, the required back pressure becomes excessive as the bank is increased. Therefore, the instructor should encourage the use of enough elevator trim to maintain a level 30° banked turn without excessive elevator control pressures during the steep turn entry or recovery.

Common student errors during steep turns include improper pitch, bank and power coordination during roll-in and roll-out, as well as uncoordinated use of flight controls or inappropriate control applications. Another student error during recovery is improper technique in correcting altitude deviations. If the airplane begins descending during the steep turn, the instructor should explain that increasing back pressure to stop the sink rate does nothing more than increase the wing loading and stall speed and, ultimately, tightens the turn. The student should shallow the angle of bank, stop the rate-of-descent, attempt to regain the altitude, *then* reestablish the desired angle of bank for the turn. Other errors include loss of orientation and excessive deviation from the desired roll-out heading.

EMERGENCY PROCEDURES

The introduction to emergency procedures should begin after the student has mastered multi-engine airwork and traffic pattern procedures and is completely familiar with the handling characteristics of the airplane. The emergency procedures associated with the multi-engine rating include engine-out operations and maneuvers and systems and equipment malfunctions.

ENGINE-OUT PROCEDURES AND MANEUVERS

Prior to introducing the engine-out procedures and maneuvers, the instructor should conduct a thorough ground training session on the techniques and theories of flight with one engine inoperative. It should be obvious that the student's success is totally dependent on an understanding of the material prior to the flight lesson. The briefing should include not only the subject matter of the lesson but also the aerodynamics of engine-out operations, the handling characteristics of the airplane, and engine-out performance considerations.

INTRODUCTION TO ENGINE-OUT OPERATIONS

It is important that the student's first experience with engine-out flight *not* contain an element of surprise, since surprise very commonly creates anxiety and tension that continues throughout the remaining lessons, retarding the learning process. To begin, the instructor should require the student to first establish the airplane in straight-and-level flight at the cruise airspeed and power setting. With the student controlling the airplane, the instructor should *slowly* close one throttle while the student applies the appropriate rudder to maintain a constant heading. In this way, the student has the opportunity to gradually apply the necessary rudder and observe the effects of decaying directional control as power is reduced. This condition also affords the instructor the opportunity to emphasize the indications associated with proper identification of the inoperative engine. Specifically, the student should understand that the first indication of an engine failure is the resultant yaw and roll, which is *always in the direction of the inoperative engine*. In addition, the control application necessary to stop yaw provides a second method of determining the inoperative engine because the foot which is *not* applying rudder pressure is on the same side as the inoperative engine.

Once the student understands the characteristics of the airplane, level, shallow banked turns with the propeller windmilling should be executed in both directions to determine the control pressure required and to gain familiarity with the handling characteristics of the airplane with one engine inoperative. To further expand upon the demonstration, the student should extend the landing gear and use various increments of flap to show the effects of drag on airspeed and the airplane's capability to climb and/or maintain altitude. The entire demonstration should be centered around the effects of the windmilling propeller, landing gear, and flaps on the performance of the airplane.

SIMULATED ENGINE-OUT MANEUVERING

Following the demonstration of drag relationships, the student should have the opportunity to maneuver the airplane with only one engine producing power. This is most easily accomplished by requiring the student to retract the landing gear and flaps and establish the zero thrust power setting published by the airplane manufacturer. In this configuration the student should begin practice with engine-out climbs, descents, and climbing and descending turns, including the use of appropriate trim. The important aspect of this instructional unit is that the student must learn not only to control airspeed, but learn the significance and the use of V_{YSE}. Specifically, *any airspeed above or below V_{YSE} reduces engine-out climb performance*. In addition, if the airplane is not capable of maintaining altitude with one engine inoperative, the use of V_{YSE} produces the *minimum rate of descent*. A detailed analysis of these concepts is discussed later in this chapter when the aerodynamic factors of engine-out flight are reviewed.

The instructor should encourage the use of a predetermined rate of descent (not to exceed 500 feet per minute) at a constant airspeed during the instruction of engine-out descents. This technique requires the student to reduce power on the operating engine rather than simply lowering the pitch altitude. This instructional technique serves two purposes.

First, the student learns how much power must be reduced to achieve the desired rate-of-descent without an increase in airspeed. Second, the student learns at what point the power must be added to stop the sink rate to allow level-off at the predetermined altitude. In addition, it is beneficial to have the student extend the landing gear using various increments of flap during the descent to demonstrate the effect on level-off power requirements. Although these exercises seem basic, engine-out traffic pattern entries, approaches, and landings are less difficult to perfect if the student masters these procedures.

ENGINE-OUT PROCEDURES

When the student no longer has difficulty controlling the airplane with only one engine producing power, the instructor should introduce the actual procedures

that are used in the event one engine loses power. It is important that the procedures be executed in strict accordance with the manufacturer's recommendations published in the pilot's operating handbook. In addition, the instructor must emphasize that the procedure should be executed in an unhurried, methodical, and accurate manner. Any unnecessary haste could result in a definite safety compromise. To insure the procedures are correctly executed, the student must be familiar with the content of the required checklists. Many instructors find it beneficial to require the student to commit the engine shutdown checklist to memory prior to the flight lesson concerning the procedure.

The procedures required for engine-out operation are most beneficially taught if the student first understands the philosophy of the emergency checklist. The instructor should stress that the engine-out emergency checklist ultimately provides for maximum available power, the elimination of drag, and the securing of the engine. These three items are prerequisites to the successful continuation of single-engine flight. To expand upon this philosophy, the instructor should analyze the checklist and teach the student the correct sequence for completing the items. A simplified engine shutdown checklist is shown in figure 5-5. Many manufacturer's recommend that engine-out procedures on checklists be committed to memory.

Engine Shutdown Procedure

To exemplify the shutdown procedures and to further analyze the checklist, the instructor must stress that the student should first maintain control of the airplane at an airspeed in excess of V_{YSE}. At the same time control of the airplane is being maintained, the mixtures should be advanced to *full rich*, the propellers moved to *high RPM*, and *both* throttles increased to *maximum power*. This procedure insures maximum power from the operating engine, regardless of which engine is inoperative.

ENGINE SHUTDOWN AND SECURING PROCEDURES

1. Maintain directional control by applying rudder to maintain a constant heading.
2. Maintain V_{YSE} as a minimum airspeed.
3. Apply maximum power by moving the mixtures, props, and throttles full forward.
4. Reduce the drag by retracting the flaps and landing gear.
5. Check auxiliary fuel boost pumps—ON.
6. Check that the fuel selector is in the proper position.
7. Identify the inoperative engine by using the "idle foot, idle engine" test.
8. Verify the inoperative engine by closing the appropriate throttle.
9. Feather the propeller by moving the propeller lever to the feather position.
10. Move the mixture control to idle cutoff.
11. Trim excess control pressures.
12. Secure the inoperative engine by using the following steps.
    - A. Fuel selector—OFF
    - B. Auxiliary boost pump—OFF
    - C. Magneto switches—OFF
    - D. Alternator—OFF
    - E. Cowl Flap—CLOSED
13. Review checklist.

Fig. 5-5. Engine Shutdown Checklist

After the application of maximum power, the auxiliary boost pumps should be activated and the fuel selector checked for selection of the proper tank. Next, the performance of the airplane should be increased by retracting the landing gear and flaps to eliminate drag. At this point in the checklist the operative engine will be producing maximum power. Now, the student can devote attention to first identifying, then verifying, and finally shutting down the inoperative engine.

Although the student should have an accurate indication of which engine has failed by the direction of yaw and rudder displacement, the instructor should emphasize the need for verifying which engine is inoperative. Generally, this verification is accomplished by retarding the throttle of the suspected failed engine. The student should understand that, if the correct engine has been selected and has completely failed, there will be no change in performance or the amount of rudder pressure needed

to counteract yaw. However, if the engine is developing partial power, the asymmetrical thrust condition will worsen. If the incorrect engine is selected, the asymmetrical thrust will decrease.

After this final confirmation, the student can continue with the propeller feathering, engine shutdown, and engine securing procedures. This normally entails placing the propeller lever in the feather position and the mixture control in idle cut-off. To secure the engine, the magnetos, alternator, and boost pump on the inoperative engine are turned off, the cowl flaps closed, and electrical load reduced for single alternator operation. The checklist is complete at this point, and a landing should be planned and accomplished as soon as practical.

Shutdown Ramifications

In addition to the shutdown of the inoperative engine, the student must be reminded to give close attention to the operating engine from two standpoints. First, the combination of reduced airspeed and additional power demanded from the engine may cause overheating problems. Therefore, the oil temperature, oil pressure, and cylinder head temperature must be monitored and maintained in the normal operating ranges. If necessary, the cowl flaps should be opened and the mixture enriched to aid in engine temperature reduction. Second, single-engine flight also creates single alternator operation, which requires an appropriate reduction in electrical load. In addition, the student must understand fully that other systems such as hydraulics and pneumatics may have greatly reduced capabilities or become completely inoperative.

Engine Restart

A common occurrence during multi-engine flight training is the propeller unfeathering and engine restart procedure. Figure 5-6 shows a typical checklist used for this procedure. It must be explained to the student why it is necessary to start propeller rotation with the engine starter. Namely, it is necessary to create oil pressure from the propeller governor, which begins to move the propeller blades from high pitch (feather) to low pitch. The starter can be disengaged as the blades move out of the feather position, because the airflow passing over the blades will continue the rotation.

The student also should be cautioned that, once the engine starts, the power should be kept to a minimum until the cylinder head and oil temperatures are within the normal operating ranges. This is important because the use of high power settings on a cold engine can result in engine damage.

ENGINE-OUT MANEUVERS

When the student understands engine-out emergency procedures, the instruc-

---

RESTARTING PROCEDURES

1. Magneto Switches—ON
2. Fuel Selector—MAIN TANK
3. Throttle—FORWARD
   approximately one inch
4. Mixture—AS REQUIRED
   for flight altitude
5. Propeller—FORWARD of detent
6. Starter Button—PRESS—RELEASE
   when engine fires
7. Mixture—AS REQUIRED
8. Power—INCREASE after cylinder head
   temperature is adequate
9. Cowl Flap—AS REQUIRED
10. Alternator—ON

Fig. 5-6. In-Flight Engine Restart Procedure

tor can introduce unannounced simulated engine failures at altitude. As the student continues to gain proficiency, simulated engine failure during takeoffs, approaches, and landings can be introduced.

SIMULATED ENGINE-OUT ON TAKEOFF

Simulated engine failures during takeoff should be regarded as a maneuver that requires a great deal of instructor discretion and judgment because the student's reaction may be totally unpredictable. For this reason, the instructor must carefully plan the introduction and notify the student of the impending failure. Later in the training, unannounced engine-out operations may be appropriate, but the requirement for instructor judgment remains unchanged. For example, if the instructor plans to introduce a simulated engine failure during the takeoff ground run, the similated failure is as instructionally effective 20 knots below rotation speed as it is at the rotation speed. Therefore, there is no reason to compromise safety for instructional purposes.

The same instructional technique should be used when dealing with a simulated power loss after the airplane is airborne during the departure climb. The instructor should first inform the student of the point at which the simulated power loss will occur. After the student demonstrates the ability to safely control the airplane and maintain the proper airspeed throughout the power loss and simulated engine shut-down, the instructor can introduce un-announced simulated power losses. The point at which the power loss occurs must be selected with discretion. For example, the instructor should not reduce power on any engine until the landing gear is in transit and the airplane has reached V_{YSE}. If it is necessary to practice engine-out procedures below V_{YSE}, the maneuver should be taught at a safe altitude rather than in the traffic pattern.

The student must realize that the specific procedure to follow after an engine failure depends on the situation. Any indication of engine power loss while the airplane is still on the ground is an automatic "no-go" situation. However, if airborne when the failure occurs, the factors affecting the pilot's decision to continue flight are many. In their POHs, some manufactur-ers use a concept called "area of decision" for applying takeoff planning in light twins. The area of decision begins where the aircraft lifts off and ends where it has reached obstacle clearance altitude. Al-though the aircraft passes through the area of decision in a few seconds, the stu-dent must be prepared to decide whether to continue or discontinue the takeoff should an engine fail. An experienced pilot usually makes this decision prior to begin-ning the takeoff roll by considering such variables as runway length and gradient, accelerate-stop distance, obstacle height in the area, takeoff weight, single-engine performance, density altitude, and level of proficiency. With an awareness of the vari-ables, the student should be taught to formulate a takeoff plan appropriate to the conditions so immediate action can be taken if an engine fails during the takeoff process. The following is an example of the type of planning the student should do for each takeoff. This example cannot be applied to every takeoff, since condi-tions may vary significantly.

1. If the engine failure occurs prior to lift-off speed, the throttles should be closed and the airplane stopped.
2. If the engine failure occurs after ob-taining obstacle clearance speed, but prior to gear retraction, the throttles should be closed and the airplane landed.
3. If the engine failure occurs above obstacle clearance speed, during or after gear retraction, maintain at least V_{XSE} (preferably V_{YSE}), feather the propeller on the inoperative engine, and climb or settle straight ahead under control.

SIMULATED ENGINE-OUT APPROACHES AND LANDINGS

The primary objective of the simulated engine-out approach and landing is to teach the student to execute a well-planned approach so an engine-out go around is unnecessary. The student should be well aware that, because of density altitude and typical engine-out performance in light twins, a successful go-around may be questionable and, in many situations, impossible.

The engine-out approach and landing should be taught in much the same man-ner that the normal, multi-engine ap-proach is taught. Specifically, the instruc-tor should stress the importance of a "plan of action" which results in an accurate, stabilized approach. It is best to lead the turns in the pattern to minimize bank angle. Other than that, the engine-out traffic pattern should be flown about the same as a multi-engine pattern with refer-ence to size and altitudes. One of the major differences with one engine inoper-ative, is that flap and landing gear exten-sion are delayed until the landing is assured.

To begin the maneuver, the pilot's operat-ing handbook should be consulted to deter-mine the recommended engine-out ap-proach speed. However, under no circum-stances should this airspeed be less than

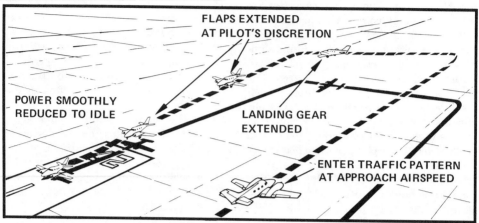

FLAPS EXTENDED
AT PILOT'S DISCRETION

POWER SMOOTHLY
REDUCED TO IDLE

LANDING GEAR
EXTENDED

ENTER TRAFFIC PATTERN
AT APPROACH AIRSPEED

Fig. 5-7. Engine-Out Approach to Landing

the single-engine best rate-of-climb airspeed. Once the proper airspeed is determined, the student should plan to enter the traffic pattern at that airspeed in level flight, as shown in figure 5-7. Then, at the preplanned point within the traffic pattern, the power should be reduced to produce a stabilized 500-foot-per-minute descent. It is important that the student knows the approximate power setting that will result in the desired rate-of-descent. If this technique is used, the student will not make excessive power changes that will require a correction later in the pattern.

During the later stages of the base leg or the final turn, the landing gear is extended and the necessary power added to maintain the required descent profile and airspeed to the point of intended landing. At the pilot's discretion, the flaps may be extended on short-final to slow the touchdown speed. However, the gear and flaps generally should not be extended until a landing is assured. After the airplane is safely on the landing roll, the instructor should emphasize the importance of directional control under conditions of asymmetrical drag.

Private and commercial applicants for multi-engine ratings who are instrument rated and who want to exercise the privileges of the instrument rating in the multi-engine airplane, are required to demonstrate competency in instrument flight.

The required instrument procedures include simulated engine failure during straight-and-level flight and turns, an instrument approach with all engines operating, and an instrument approach that simulates one engine inoperative. Instructors should refer to the multi-engine section of the private or commercial PTS, as appropriate, for performance criteria. Multi-engine rating applicants who do not wish to demonstrate instrument competency must say so at the beginning of the practical test. Their privileges will be limited to "VFR only" in the multi-engine airplane.

LOSS OF DIRECTIONAL CONTROL DEMONSTRATION

The engine inoperative loss of directional control demonstration should be accomplished within a safe distance of a suitable airport and at an altitude no lower than 3,000 feet AGL. To begin, the propellers are set to high r.p.m., the landing gear is retracted, and the wing flaps, cowl flaps, and trim are set to the takeoff position. Then, the aircraft is placed in a climb attitude with airspeed established similar to that following takeoff. Excessive nose-high pitch attitudes should be avoided. Next, with both engines developing rated takeoff power, or as recommended, power on the critical engine (usually the left) is reduced to idle (windmilling, not shut down). Abrupt power reduction should be avoided. After this, a bank toward the operating engine is established, as necessary, for best performance. Airspeed is *slowly* reduced with elevators while using rudder pressure to maintain directional control until all available rudder is applied. At this point, recovery should be initiated by simultaneously reducing power on the *operating* engine and by lowering the nose to decrease the angle of attack. Recovery *should not* be made by increasing power on the simulated inoperative engine. The instructor should emphasize the importance of conserving altitude during recovery, *but not at the expense of uncontrolled flight.* If necessary, power on the operating engine can be reduced to maintain control with minimum loss of altitude.

One demonstration should be made while holding the wings level and the ball centered, and another demonstration should be made while banking the airplane at least 5° toward the operating engine to establish "zero side slip." These maneuvers will demonstrate the engine-out minimum control speed for the existing conditions and will emphasize the necessity of banking into the operative engine.

If indications of a stall occur prior to loss of directional control, recovery should be started immediately by reducing the angle of attack. In this case, an engine-out minimum control speed demonstration is not possible under the existing conditions.

Instructors should stress that there is a certain density altitude above which the stalling speed is higher than V_{MC}. When this condition exists close to the ground because of high elevations or temperature, the loss of directional control demonstration should not be attempted. In this event, the instructor should emphasize the significance of engine-out minimum control speed, including the possible consequences of flight below this speed, cues of a pending loss of control, and the recovery technique.

Common student errors related to loss of directional control are contained in the following list.

1. Inadequate knowledge of the causes of loss of directional control at airspeeds less than V_{MC}, factors affecting V_{MC}, and safe recovery procedures
2. Improper entry procedures, including pitch attitude, bank attitude, and airspeed
3. Failure to recognize imminent loss of directional control
4. Rough and/or uncoordinated control technique

A review of the aerodynamic factors that determine V_{MC} also is recommended. For example, in normally aspirated engines, actual V_{MC} decreases with altitude because power decreases with altitude. In addition, actual V_{MC} is higher when the CG is at the aft limit. This is illustrated in figure 5-8 by the decreased length of the arm from the aft CG to the rudder.

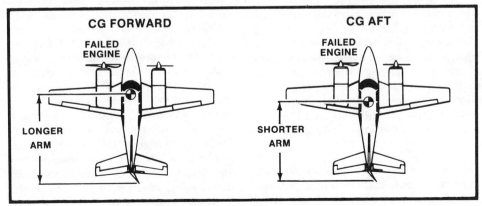

Fig. 5-8. Aft CG Increases V_{MC}

SECTION B—AERODYNAMICS OF MULTI-ENGINE FLIGHT

A solid basis in the principles of multi-engine aerodynamics will allow the instructor to maintain a high degree of safety during multi-engine training. In addition, students who are familiar with the subject of aerodynamics tend to learn faster during training and quickly develop judgment concerning the safety of multi-engine operations. Instructors will find that explaining the application of aerodynamic principles to the student during training will be time well spent.

BASIC PRINCIPLES

The principles of aerodynamics governing the flight of single-engine airplanes also apply to those with more than one engine. Multi-engine airplanes respond to the four forces of flight just as any other airplane, regardless of the number of powerplants. The primary aerodynamic differences between single-engine and multi-engine flight result from the location of the engines.

Thrust in conventional multi-engine airplanes is not directed along the centerline as it is in airplanes with one engine. Conventional twin-engine airplanes have one engine mounted on each wing, producing thrust *parallel* to the longitudinal axis. Therefore, the thrust vectors are displaced from the airplane's centerline.

INDUCED AIRFLOW

The most important advantage of wing-mounted engines is that a substantial amount of lift is derived from propeller slipstream. This *induced airflow* also occurs in single-engine airplanes; however, it is not as effective or apparent in aircraft where the powerplant is mounted along the centerline. Figure 5-9 illustrates the effect of induced airflow.

The lift provided by induced flow is an important consideration during landing and in imminent stall recoveries. A sharp power reduction will cause an instantan-

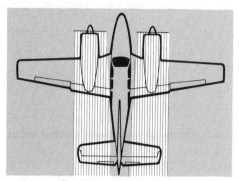

Fig. 5-9. Induced Airflow

eous loss of propeller slipstream flow and a corresponding loss of total lift. This could result in a high sink rate or stall if power is reduced suddenly during a slow final approach. On the other hand, rapid addition of power when an imminent stall has been identified could prevent further stall development. However, induced flow should never be relied on as the sole factor in stall recovery. The addition of power in combination with other normal stall recovery procedures should be utilized.

TURNING TENDENCIES

In single-engine airplanes, left-turning tendencies are caused by both asymmetrical propeller loading (P-factor) and torque. Multi-engine airplanes have an even greater tendency to turn during climbs or other high angle of attack maneuvers due to the additional engine and propeller. The engine positions in relation to the airplane centerline causes asymmetrical propeller loading to exert a forceful turning moment, as shown in figure 5-10. In addition, the effects of torque also are greater in multi-engine airplanes because of the more powerful engines usually encountered. Torque is caused by the tendency of the airplane itself to turn about the propeller and engine crankshaft. It acts in the direction opposite propeller rotation, as shown in the top portion of figure 5-11. In twins equipped with counterrotating engines,

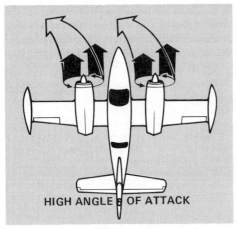

Fig. 5-10. Asymmetrical Propeller Loading

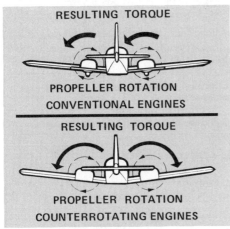

Fig. 5-11. Conventional vs. Counterrotating Engines

however, the effect of torque is eliminated. The torque from one engine cancels the effect of the other, as shown in the lower portion of figure 5-11.

MULTI-ENGINE CEILINGS

Normally aspirated engines progressively lose power as the airplane ascends because of decreasing atmospheric pressure. This gradual loss of power determines the maximum altitude capabilities of the airplane. Contrary to popular belief, climb performance is based on excess power and not on the creation of additional lift. The amount of lift produced during a climb is nearly the same as the amount required for level flight.

Turbocharging considerably increases the operating altitudes over normally aspirated engines, by providing sea level or greater manifold pressure at high altitudes where normally aspirated engines can only provide manifold pressure equal to the ambient atmospheric pressure. However, once the airplane has climbed to the turbocharger's critical altitude, the engines begin to progressively lose power with altitude. From that point, turbocharged engines progressively lose power like their normally aspirated counterparts.

SERVICE CEILING

The *service ceiling* is defined as the maximum density altitude where the best rate-of-climb airspeed will produce a 100 f.p.m. climb at maximum weight in the clean configuration with both engines producing maximum continuous power. The *single-engine* service ceiling refers to the altitude at which the maximum rate of climb is only 50 f.p.m. with one engine inoperative.

ABSOLUTE CEILING

The *absolute ceiling* is defined as the maximum density altitude the airplane is capable of attaining or maintaining at maximum weight in the clean configuration and at maximum continuous power. The absolute ceiling is closely related to the best angle-of-climb and best rate-of-climb speeds. As altitude increases, the best angle-of-climb speed increases, while the best rate-of-climb speed decreases. The point where the two speeds converge is the absolute ceiling.

MULTI-ENGINE V-SPEEDS

The chart in figure 5-12 lists the critical airspeeds (V-speeds) normally encountered in multi-engine flying. The manufacter's operating instructions should be consulted for other speeds, such as the operating range for landing gear and wing flaps.

ENGINE-OUT AERODYNAMICS

Engine-out procedures and techniques are a critical part of multi-engine instruc-

Designation	Description	Airplane Configuration or Significance of Speed	A/S Ind. Marking
V_{SO}	Stalling Speed—Landing Configuration	Engines zero thrust, propellers takeoff position, landing gear extended, flaps in landing position, cowl flaps closed	Low speed end of white arc
V_{S1}	Stalling Speed—Specified Configuration	Engines zero thrust, propellers takeoff position, landing gear and flaps retracted	Low speed end of green arc
V_{MC} V_{MCA}	Minimum Control Airspeed (Air Minimum Control Speed)	Takeoff or maximum available power on operating engine, critical engine windmilling (or feathered if auto feather device is installed), landing gear retracted, flaps in takeoff position, and CG at aft limit	Red radial line *
V_{SSE}	Intentional One Engine Inoperative Speed	Minimum speed for intentionally rendering one engine inoperative in flight for pilot training	
V_X	Best Angle-of-Climb Speed	Speed which produces most altitude gain over a given distance with both engines operating; obstruction clearance speed	
V_Y	Best Rate-of-Climb Speed	Speed which produces most altitude gain in a given time with both engines operating	
V_{XSE}	Best Angle-of-Climb Speed (Single Engine)	Speed which produces most altitude gain over a given distance with one engine inoperative	
V_{YSE}	Best Rate-of-Climb Speed (Single Engine)	Speed which produces most altitude gain in a given time with one engine inoperative	Blue radial line *
V_{LO}	Maximum Landing Gear Operating Speed	Maximum speed for safely extending or retracting the landing gear	
V_{LE}	Maximum Landing Gear Extended Speed	Maximum speed for safe flight with landing gear extended	
V_{FE}	Maximum Flap Extended Speed	Maximum speed with wing flaps in a prescribed extended position	High speed end of white arc
V_A	Design Maneuvering Speed	Speed below which structural damage will not occur as a result of full control deflection	
V_{NO}	Maximum Structural Cruising Speed	Maximum speed for normal operation	High speed end of green arc
V_{NE}	Never-Exceed Speed	Maximum design speed without structural failure	Red line

* Aircraft receiving type certificates under FAR Part 23 after November 11, 1965

Fig. 5-12. Multi-Engine V-Speeds

tion. A safe and proficient pilot must thoroughly understand the aerodynamic factors involved in engine-out flight. The instructor also must have a complete understanding of these factors to insure safety and the proper training of each student.

When an engine fails, the greatest overall performance loss results from a 50 percent reduction in available horsepower. However, the total loss of *climb performance* when one engine fails is approximately 80 percent for most light, conventional, twin-engine airplanes. Since climb performance is based on excess power available, an engine failure reduces excess power by more than one-half because of the sharp increase in drag that the remaining engine's power must overcome. The sudden drag increase results from asymmetrical thrust and the control responses needed to counter yawing and rolling tendencies. This means a multi-engine airplane may not climb at all with one engine inoperative. Although aircraft certified under FAR Part 23 are required to meet certain performance and handling criteria, many light twin-engine airplanes are not required to demonstrate climbing ability during single-engine operations. The regulation requires a positive engine-out rate of climb at 5,000 feet density altitude only for those multi-engine airplanes weighing more than 6,000 pounds and/or having a stalling speed (V_{SO}) greater than 61 knots. If the airplane weighs 6,000 pounds or less and has a stalling speed of 61 knots or less, a specific, *positive* engine-out rate of climb is not required. However, the manufacturer must determine what the engine-out climb performance is at 5,000 feet with the critical engine inoperative and its propeller in the minimum drag configuration.

CRITICAL ENGINE

The term *critical engine* refers to the engine whose failure most adversely affects the performance or handling characteristics of the airplane. Consideration of the critical engine is most significant in situations where the airplane is operating at low speeds with a high power setting. In these high angle of attack attitudes, the descending blade of each propeller is producing more thrust than the ascending blade. The unequal thrust vector can be related to the "moment" presented in weight and balance. The longitudinal axis

of the airplane is considered the datum for this discussion, and each of the descending propeller blades is given an "arm." Thus, the thrust from each engine may be measured as a "moment arm," as illustrated in figure 5-13.

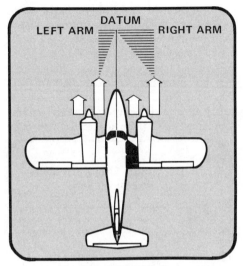

Fig. 5-13. Critical Engine

If the left engine failed, directional control would be affected more than if the right engine became inoperative. the larger moment arm associated with the right engine would have a greater influence than that of the left engine. Therefore, in conventional multi-engine airplanes, the left engine is the critical engine.

MINIMUM CONTROL SPEED

Each multi-engine airplane certified under FAR Part 23 has an established minimum control speed (V_{MC}) with the critical engine inoperative and the other engine developing takeoff power.

V_{MC} is established under specified criteria including conditions that affect controllability. The airplane is in the most critical takeoff configuration with the propeller controls in the recommended takeoff position and the landing gear retracted. The airplane must also be loaded to maximum takeoff weight (or any lesser weight necessary to show V_{MC}) with the most unfavorable center of gravity. Additional factors include the airplane trimmed for takeoff and airborne with ground effect negligible. Finally, a bank angle of not more than five degrees is established toward the good engine.

The established V_{MC} is found in the pilot's operating handbook and is normally shown on the airspeed indicator with a red radial line. However, the actual V_{MC} varies as any of the specified conditions change. For example, moving the CG forward will decrease V_{MC}. In this case, the rudder is more effective in counteracting the yawing tendency due to the longer arm through

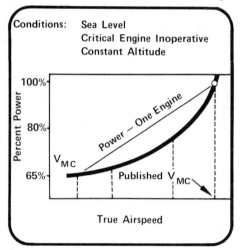

Fig. 5-14. Percent Power and Minimum Control Speed

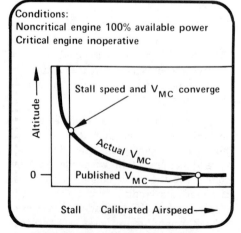

Fig. 5-15. Effect of Altitude on V_{MC}

which it works. Another variable is the power output of the operating engine. If power is decreased, the tendency for the airplane to yaw and roll toward the inoperative engine is also decreased, lowering V_{MC}. Figure 5-14 illustrates that there is a wide range of actual minimum control airspeeds with different power settings. In addition, as altitude increases, normally aspirated engines lose efficiency and are unable to develop 100 percent rated sea level power. This power loss also causes actual V_{MC} to decrease.

While actual V_{MC} is decreasing with altitude, the calibrated stalling speed remains the same. Figure 5-15 illustrates the effect of altitude on actual V_{MC} and shows that stalling speed and V_{MC} eventually converge. Therefore, safe flight below published V_{MC} should not be assumed when at high density altitudes.

WEIGHT AND CENTER OF GRAVITY

Data obtained from full-scale wind tunnel tests of a light twin aircraft show that aircraft weight directly affects V_{MC}. The tests show that as weight decreases, actual V_{MC} increases. The wind tunnel tests also confirm that as the CG shifts forward, V_{MC} decreases. Typically, the CG has a tendency to shift forward with fewer passengers, but cargo is often loaded in the aft baggage areas. When flying with few passengers and a lot of cargo, it may be advantageous to load the seats (if cargo can be properly secured) so the CG will be toward the forward side of the center of gravity envelope.

STABILITY AND CONTROL

The primary concern after engine failure is to keep the aircraft controllable and stabilize the attitude for the best performance. Maintaining V_{MC} does not guarantee that the airplane can climb or even hold altitude, but only that the pilot can maintain a heading. Control effectiveness in any flight situation depends on the velocity of the airflow over the control surfaces. At airspeeds below V_{MC}, full control deflection does not provide sufficient force to overcome the yawing and rolling tendencies induced by failure of the critical engine.

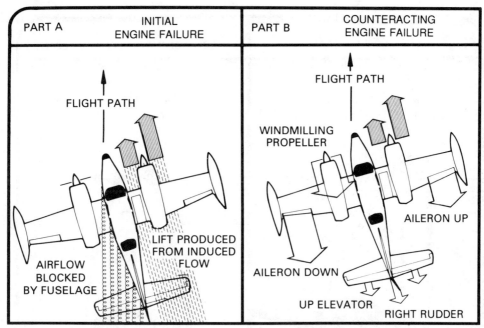

Fig. 5-16. Asymmetrical Forces

IN-FLIGHT FORCES

During flight with one engine inoperative in a multi-engine airplane, the four forces take on new dimensions. Three of them (lift, thrust, and drag) now act on the aircraft asymmetrically, while the management of the fourth (weight) is more critical.

When either engine fails, the induced airflow from the propeller slipstream is lost and total lift on that wing is decreased (excluding airplanes with a centerline thrust configuration). However, the induced airflow from the operating engine is still producing additional lift, and this lift differential causes a rolling tendency toward the inoperative engine as shown in figure 5-16, Part A.

Figure 5-16 also illustrates the yawing tendency toward the inoperative engine caused by the asymmetrical thrust condition. This situation causes some of the airflow to be blocked by the fuselage since the flight path is no longer parallel to the longitudinal axis.

When control surfaces, such as the ailerons or the rudder, are displaced from the neutral position, a proportional amount of drag is created. Figure 5-16, Part B illustrates this induced drag as well as the parasite drag created by the propeller of the inoperative engine. Depending on the blade angle, the drag of a windmilling propeller may be as severe as that of a flat disc of the same diameter. Figure 5-17 shows the relationship of the windmilling propeller and stationary propeller to blade angle and drag created. This comparison illustrates importance of feathering the propeller of the inoperative engine promptly to reduce drag, assuming it cannot be restarted immediately.

Situations should be avoided where aircraft loading could affect engine-out operations. As stated earlier, the CG position affects controllability, and loading the airplane to the aft limit increases V_{MC}. In airplanes equipped with wing-mounted tanks outboard of the engines, fuel management may also be critical. For example, a heavy fuel load on the side of the critical engine and a light fuel load on the other side should be avoided. As always, loading should follow the recommendations in the pilot's operating handbook.

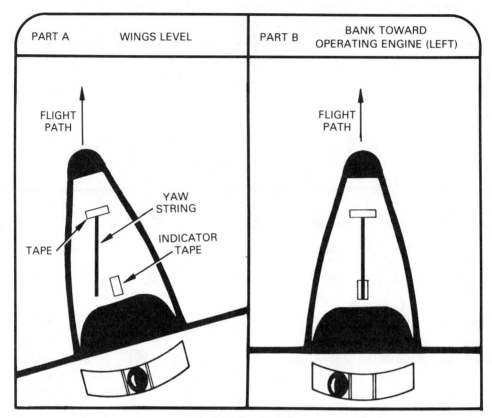

Fig. 5-18. Zero Sideslip

CONTROL APPLICATION

Considerable control pressures are necessary to counter the rolling and yawing moments encountered during flight with one engine. However, rudder deflection can be minimized if the airplane is banked up to 5° toward the operative engine. In this configuration, the high wing produces a lift vector which acts to relieve the rudder forces required to maintain straight flight. A common misconception in engine-out operations is that the airplane should be flown with the wings level and the ball centered in the inclinometer, as shown in figure 5-18, Part A. This causes a side slip into the inoperative engine and can be illustrated to the student by attaching a yaw string to the airplane. Flight tests have shown that this can increase actual V_{MC} by 20 knots in some airplanes and decrease single-engine climb performance by several hundred feet per minute. Banking toward the operative engine reduces V_{MC} and ensures that stall characteristics will not be degraded. Using enough rudder to maintain straight flight while maintaining up to 5° of bank puts the airplane in a zero side slip condition. The ball will be about one ball width toward the "good" side, as shown in figure 5-18, Part B.

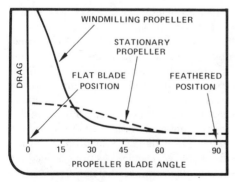

Fig. 5-17. Propeller Drag Contribution

COUNTERROTATING ENGINES

When an airplane is equipped with counterrotating propellers, neither engine is considered critical. Turning tendencies are still very apparent, but lack the severity of those associated with the critical engine failure of a conventional twin.

ENGINE-OUT V-SPEEDS

An engine failure in a twin usually means the best climb performance available is only marginal when compared to the performance with both engines operating. Climb airspeeds are published by the manufacturer for best rate and best angle during single-engine operation. These airspeeds are referred to as V_{YSE} and V_{XSE}, respectively.

BEST SINGLE-ENGINE RATE OF CLIMB

V_{YSE} is the single-engine best rate-of-climb speed which produces the greatest gain in altitude in a given amount of time with one engine inoperative. This airspeed normally is used if altitude is a prime consideration after engine failure. If engine failure occurs above the airplane's single-engine service ceiling, this airspeed produces the minimum descent rate with the operative engine at a maximum continuous power.

BEST SINGLE-ENGINE ANGLE OF CLIMB

V_{XSE} is the engine-out best angle-of-climb airspeed. This speed is used for obstruction clearance with one engine inoperative. In most cases, V_{XSE} is a somewhat higher airspeed than the twin-engine best angle-of-climb speed (V_X).

SINGLE-ENGINE CEILINGS

The operating altitudes of multi-engine airplanes are lowered considerably during single-engine flight. The engine loss reduces total thrust available and corresponding reductions of *service ceiling* and *absolute ceiling* result.

SINGLE-ENGINE SERVICE CEILING

The single-engine service ceiling is the maximum density altitude at which the single-engine best rate-of-climb airspeed (V_{YSE}) will produce a 50 f.p.m. rate of climb. The ability to climb 50 f.p.m. in calm air is necessary to maintain *level flight* for long periods in turbulent air. This ceiling assumes the airplane is at maximum weight in the clean configuration, the critical engine (if appropriate) is inoperative, and the propeller is feathered. In comparison, the multi-engine service ceiling is the density altitude at which the best rate-of-climb airspeed (V_Y) will produce a 100 f.p.m. rate of climb at maximum weight in the clean configuration.

SINGLE-ENGINE ABSOLUTE CEILING

The single-engine absolute ceiling is the maximum density altitude the airplane is capable of attaining or maintaining. This ceiling assumes the airplane is at maximum weight in the clean configuration, the critical engine (if appropriate) is inoperative, and the propeller is feathered. This is also the density altitude at which V_{XSE} and V_{YSE} are the same airspeed.

Figure 5-19 illustrates an engine failure at a cruising altitude above the single-engine absolute ceiling. In this situation, the aircraft will descend gradually (drift down) to the density altitude equivalent of the single-engine absolute ceiling appropriate to the existing airplane weight. In turbulent air, the airplane may not be capable of maintaining the single-engine absolute ceiling and may continue to descend to the single-engine service ceiling or lower.

SINGLE-ENGINE CLIMB

Climb performance is dependent upon the ratio of thrust to drag. If drag exceeds thrust, climb is not possible and the best rate-of-climb speed will provide the least rate of descent. Figure 5-20 depicts a typical drag curve with available thrust when one engine is inoperative. Any thrust and airspeed combination that falls within the shaded area represents a positive rate of climb. Conversely, any point outside the

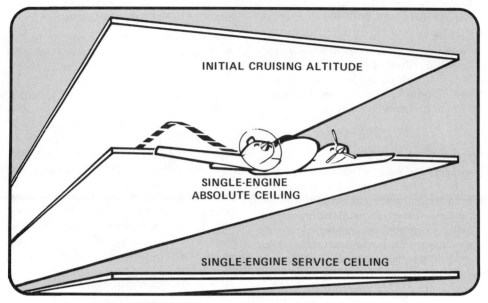

Fig. 5-19. Altitude Loss During Single-Engine Operation

shaded area will produce a descent. This graph also illustrates the thrust available on the same airplane with one engine at 15,000 feet. At that altitude, drag exceeds thrust and a positive rate of climb cannot be established.

Drag is a major factor in the amount of excess thrust available. Any increase in drag, whether parasite or induced, must be balanced by the use of additional thrust. Any use of excess thrust to counteract drag is thrust which cannot be used for the climb. Since the extension of the landing gear and/or flaps increases parasite drag, the maximum amount of excess thrust is available in the clean configuration. In most light and medium twin-engine airplanes, the extension of the landing gear or flaps may increase drag in excess of the available thrust causing a descent.

Drag also increases because of the rudder and aileron deflections necessary to maintain straight flight and the airflow restriction caused by the inoperative engine's propeller. An increase in airspeed also increases drag. Therefore, any speed above that necessary to maintain level flight, or

above V_{YSE} during a climb, requires excessive thrust and decreases the airplane's performance. All of this results in a dramatic decrease in performance. It is important that the student be familiar with the correct order for drag reduction following an engine failure. An extended landing gear and full flaps contribute a substantial amount of drag. A windmilling propeller also produces very significant drag as do the control deflections required to stop the airplane from turning. Since it is considered unwise to feather an engine before it has been positively identified, drag normally is reduced by first retracting full

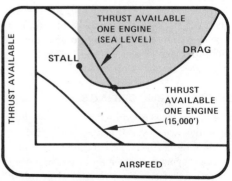

Fig. 5-20. Thrust Available for Climb

flaps and landing gear. Next, the failed engine is identified and the propeller is feathered. However, the relative amount of drag per component and the specific order of drag reduction may vary between types of twin-engine airplanes, so the manufacturer's specific recommendations should always be followed. During training, it is beneficial to use the vertical speed indicator to compare the relative engine-out performance loss associated with a windmilling propeller, full flaps, and extended landing gear.

Performance will be decreased still further by any increase in density altitude. Density altitude will affect the single-engine climb rate as it affects the twin-engine climb rate, but the results will be more critical during engine-out operations. For example, the single-engine climb rate normally is very small for most light multi-engine airplanes. Therefore, any decrease in this performance because of higher density altitude will quickly result in a negligible climb rate or a possible descent.

Climb performance can be improved greatly by reducing operating weights. As an example, a typical multi-engine airplane's sea level, single-engine climb rate may increase from 230 f.p.m. at 4,000 pounds to 420 f.p.m. at 3,400 pounds. At the same time, the single-engine service ceiling may increase from 5,200 feet to 10,000 feet under standard atmospheric conditions.

With regard to the above performance consideration, it is evident that the takeoff and climb are the most critical phases of all multi-engine operations. During this time, the airplane is the heaviest it will be throughout the flight. Therefore, the actual single-engine rate of climb and service ceiling are the lowest.

Turns, except to avoid obstructions, should be limited during a single-engine climb until the airplane has reached a safe altitude to return to the airport. Even a perfectly coordinated turn increases induced drag which, in turn, increases the amount of thrust required to maintain altitude. This results in less

excess thrust for climb. If a turn is necessary, the smallest bank angle possible for the existing conditions should be used. To illustrate this point, figure 5-21 shows that, at a constant airspeed, a 15° bank increases the induced drag by approximately seven percent, while a 30° bank results in a 33 percent increase.

Because of the increased drag and decreased control effectiveness, it is recommended that a maximum bank angle of 15° be used during single-engine maneuvering and, then, only after a safe maneuvering altitude is attained. The 15° bank should be maximum because of the rapid drag increase associated with higher bank angles. As illustrated, a five degree increase from 15° to 20° of bank nearly doubles the percentage increase in induced drag.

The entire discussion of asymmetrical thrust, V_{MC}, and single-engine climbs has pointed out that takeoff and initial climb are the most critical phases of multi-engine operations. The pilot must continually analyze the aircraft's performance in order to know what single-engine performance can be expected in the event of engine failure or partial power loss.

SPIN CONSIDERATIONS

During multi-engine flight training, the instructor should give special consideration to the situations where an inadvertent spin may develop. The following discussion is not designed to review basic spin recovery procedures but, rather, to provide background knowledge regarding multi-engine airplane type certification. In addition, specific situations are described where the post stall gyration (the first turn) could develop into a spin.

MULTI-ENGINE AIRPLANE CERTIFICATION

Since the first element of any spin is a fully developed stall, it is important to analyze the required stall characteristics for airplane certification under FAR Part 23. Before the airplane is certified, it must be flight tested and stall recovery must be demonstrated through the normal use of controls. Recovery must not produce more than 15° of roll or yaw.

In addition, FAR Part 23 requires that stalls be performed with the critical engine inoperative. During this test, the airplane must not display any undue spinning tendency and must be safely recoverable without the application of power to the inoperative engine. However, the operating engine(s) may be throttled back during the recovery.

Based on the certification criteria, the airplane should not have adverse or unpredictable spin characteristics during stall practice. However, *FAR Part 23 does not require that any multi-engine airplane be tested for spin characteristics.* As a result, the *actual characteristics* during the post-stall gyration and the spin *are unknown.* This means that spin recovery techniques are only *recommendations* and may very from the recovery techniques required for the individual situation. It is for this reason that, any time a potential spin situation is encountered, the instructor must effect an immediate recovery before the post-stall gyration can develop into a spin. If immediate corrective action is delayed,

or if incorrect recovery procedures are applied and a spin develops, recovery is *highly improbable and, in some situations, impossible.*

SPIN CONDITIONS

Based on the certification criteria for stalls, it is unlikely that a spin will develop during normal stall practice. However, the instructor must be particularly alert in the event the student uses improper recovery techniques.

It is obvious that the chances of inadvertent spin entry are greatest during the instruction of engine-out flight maneuvers. These maneuvers are particularly critical considering the relationship between V_{MC} and the single-engine stalling speed. In some density altitude situations, the airplane will stall at an airspeed higher than actual V_{MC} for the current conditions. The instructor and the student must be fully aware of this condition and begin an immediate recovery at the first indication of the stall.

The second and most critical aspect of the V_{MC} and stalling speed relationship is the situation where these two speeds are identical. In this situation, a sudden and violent loss of control may occur and an immediate recovery must be initiated before the post-stall gyration is allowed to develop into a spin.

BANK ANGLE, DEGREES	LOAD FACTOR	PERCENT INCREASE IN STALL SPEED	PERCENT INCREASE IN INDUCED DRAG (AT CONSTANT VELOCITY)
0	1.0000	0	0
5	1.0038	0.2	0.8
10	1.0154	0.7	3.1
15	1.0353	1.7	7.2
20	1.0642	3.2	13.3
25	1.1034	5.0	21.7
30	1.1547	7.5	33.3
35	1.2208	10.5	49.0
40	1.3054	14.3	70.4
45	1.4142	18.9	100.0
60	2.0000	41.4	300.0

Fig. 5-21. Bank Angle and Induced Drag

EXERCISES

INTRODUCTION

These exercises are designed to complement the *Flight Instructor Manual*. Each exercise is correlated with a specific textbook chapter and section. For example, Exercise 1A applies to textbook Chapter 1, Section A.

The exercises contain multiple choice, true/false, matching, and completion exercises. To answer the multiple choice questions, circle the number of the correct choice. Fill in the appropriate blanks to answer the other questions. Further instructions may appear at the beginning of a section or within the body of an individual exercise, when necessary. The answers to all of the exercises are grouped at the back of the manual following the Pilot Briefing section.

CHAPTER 1—THE BASIC FLIGHT INSTRUCTOR

Exercise 1A—Single-Engine Airplane Instruction

1. The term "integrated flight instruction" suggests
   1. concurrent ground and flight training.
   2. emphasis on aircraft control by visual references.
   3. emphasis on aircraft control by instrument references.
   4. control of the aircraft by both visual and instrument references.

2. Instruction in maneuvering during slow flight should stress _____ _____ .

3. The effectiveness of stall/spin awareness training can be enhanced with the use of _____ _____ .

4. _____ (True, False) A good guideline for teaching stall/spin awareness is to allocate one hour of ground and one hour of flight training.

5. The technique of placing a dot on the windshield directly in front of the student during turn practice is helpful in teaching control of _____ _____ .

6. The correct stall recovery procedure is to
   1. level the wings, add full power, and lower the pitch attitude.
   2. level the wings, lower the pitch attitude, and add full power.
   3. lower the pitch attitude, add full power, and level the wings.
   4. lower the pitch attitude and level the wings before adding full power.

7. _____ (True, False) Approximate wind direction and velocity can be determined by tracking a straight line.

8. One of the primary objectives of rectangular course and S-turn maneuvers is to teach students how to compensate for _____ .

9. When performing ground reference maneuvers that require a constant radius turn, the steepest angle of bank will occur when the airplane is positioned _____ .

10. The stabilized landing approach involves a constant power setting, _____ , and _____ _____ .

11. A student who slowly turns the aircraft to the left on the upwind leg of the traffic pattern during climbout probably is not compensating for _____ .

12. When should a student first be instructed in the use of flaps?
    1. Only after solo
    2. During cross-country flight
    3. During soft- or short-field landings
    4. From the beginning of traffic pattern training

13. _____ (True, False) Crosswind takeoffs and landings should be introduced and practiced prior to a student's first solo flight.

14. _____ (True, False) Soft- and short-field procedures are normally practiced before the student's first solo flight.

15. During a short-field or maximum performance takeoff, the aircraft should first be accelerated to the recommended liftoff speed, then a climb should be initiated at _____ until the obstacle is cleared. The climb should then be continued at _____ .

16. _____ (True, False) A soft-field takeoff requires liftoff to be accomplished at the airplane's best angle-of-climb speed.

17. The three types of navigation normally introduced during the first dual cross-country flight are _____ , _____ _____ and _____ _____ .

18. The first action a student should take following a simulated engine failure is to
    1. pick a suitable field for landing.
    2. establish the proper glide speed.
    3. apply carburetor heat.
    4. change fuel tanks.

19. During a chandelle, the highest pitch attitude should be established after turning
    1. 45°.
    2. 90°.
    3. 135°.
    4. 180°.

20. _____ (True, False) Steep power turns should be accomplished using outside visual references only.

21. _____ (True, False) During the performance of a lazy eight, the highest pitch attitude occurs at the 90° reference point.

22. _____ (True, False) During eights-along-a-road, the wind may be blowing either parallel or perpendicular to the road.

23. When the groundspeed decreases during eights-on-pylons, the pivotal altitude _____ (increases, decreases).

24. A high performance airplane is defined as one equipped with flaps, controllable propeller, and retractable landing gear, or an engine with more than
    1. 180 hp.
    2. 200 hp.
    3. 230 hp.
    4. 260 hp.

25. _____ (True, False) If a pilot fails to complete a flight review satisfactorily, an endorsement must be placed in the pilot's logbook to that effect.

Exercise 1B– Aerodynamics of Flight

1. Any imbalance in the four forces acting on an airplane in flight will cause _____ or _____ until the forces are again equalized.

2. In a constant airspeed climb, the four forces acting on an airplane are considered to be in _____ .

3. About 75% of total lift is created by a pressure differential _____ and _____ the wing.

4. About 25% of total lift is contributed by the _____ _____ of the air that is deflected downward by the wing.

5. The flow pattern around a wing is dependent on the airfoil's _____
   and _____ ____ _____ .

6. The angle of attack is the angle between the wing's _____
   _____ and the _____ _____ .

7. The point along the chord where all lift is considered to be concentrated is the
   1. center of lift.
   2. center of pressure.
   3. aerodynamic center.
   4. center of gravity.

8. The point along the chord line where all changes in lift are considered to take
   place is termed the
   1. aerodynamic center.
   2. center of pressure.
   3. center of lift.
   4. center of gravity.

9. An excessive angle of attack causes the smooth airflow over the wing to
   _____ .

10. The wingtips usually have a lower angle of attack than the wing root due to
    _____ .

11. _____ (True, False) It is desirable to have the wing stall at the tips first.

12. _____ (True, False) The higher the aspect ratio, the more efficient the wing.

13. _____ (True, False) Aspect ratio is computed by dividing the span by the
    average chord.

14. _____ (True, False) Parasite drag decreases as airspeed increases.

15. _____ (True, False) Induced drag decreases as airspeed decreases.

16. What causes induced drag?
    1. Skin friction
    2. Lift production
    3. Shape of the airplane
    4. Airflow interference

17. The main advantage of tip tanks or end plates on the wings is that they
    1. reduce induced drag.
    2. increase induced drag.
    3. reduce angle of attack.
    4. increase stall speed.

18. The airspeed at which induced drag and parasite drag are equal produces the
    1. maximum range.
    2. best economy cruise speed.
    3. maximum cruise speed.
    4. maximum endurance.

19. Ground effect alters the aerodynamic characteristics of the wing so that a given value of lift is maintained with a
    1. larger angle of incidence.
    2. smaller angle of incidence.
    3. larger angle of attack.
    4. smaller angle of attack.

20. Available thrust depends on engine shaft horsepower and propeller _____.

21. For an airplane to climb, there must exist an excess of
    1. lift.
    2. induced drag.
    3. thrust.
    4. parasite drag.

22. Rate of climb depends on the amount of excess _____ available.

23. The principal factor in propeller efficiency is its _____ _____.

24. Asymmetric loading of the propeller (P-factor) is caused by the higher angle of attack of the _____ blades.

25. Spiralling slipstream during takeoff causes the airplane to _____ to the _____.

26. The left-turning tendencies of an airplane increase with an increase in power, or _____ ___ _____.

27. _____ (True, False) The factor having the greatest effect on climb performance is increased altitude because it increases the power and thrust required but decreases the power and thrust available.

28. Best glide speed should be maintained with gear and flaps _____ (extended, retracted).

29. In a turn, the aircraft is accelerated toward the center of the turn by the _____ component of lift.

Match the definition on the right with the appropriate type of airplane stability on the left.

30. _____ Static stability

31. _____ Dynamic stability

32. _____ Longitudinal stability

33. _____ Lateral stability

34. _____ Directional stability

A. The tendency to return to a wings level attitude after a displacement

B. The tendency to remain stationary about the vertical axis.

C. The time required to return to a state of equilibrium following a displacement

D. The tendency to resist pitch displacement from level flight and to return to level flight if displaced

E. The tendency to return to a state of equilibrium after a displacement

CHAPTER 2—THEORY OF INSTRUCTION
Exercise 2A—
The Learning Process

1. _____ (True, False) A person who is convinced that learning to fly will enhance professional qualifications is an example of the law of readiness.

2. Continued practice of takeoffs and landings makes use of the law of _____.

3. Dual instruction in crosswind takeoffs and landings during moderately gusty conditions is an application of the law of _____.

4. _____ (True, False) When the instructor reviews and highlights the important elements of a class discussion, the law of recency is being exercised.

5. That which is learned first often creates an unshakable impression on the student even if the concept is wrong. This is an example of the law of
   1. exercise.
   2. effect.
   3. readiness.
   4. primacy.

6. _____ (True, False) An instructor who uses negative motivation in a learning situation is applying the law of effect.

7. The statement that, "Students do their best when anticipating a successful outcome," is an example of the law of
   1. effect.
   2. readiness.
   3. primacy.
   4. intensity

8. When perceptions about pitch attitude, power, and airspeed combine into a meaningful relationship, a student is said to have acquired _____ .

9. The desire to improve and maintain one's self-image defines a person's

   _____   _____ .

10. _____ (True, False) Negative motivation (such as threats) has a valid use in the teaching environment.

11. The objective of all instruction is to achieve the highest level of learning which is

    _____ .

12. _____ (True, False) Students who feel insecure concerning their piloting abilities must be subjected to persistent criticism to build up resistance to conflict situations.

Match the theory of forgetting on the right with the appropriate situation on the left.

13. _____ A student finds that most of   A. Disuse
    the basics learned in the first few   B. Interference
    lessons have been forgotten   C. Repression

14. _____ A student experiences diffi-
    culty remembering performance
    speeds after spending only a short
    time in another model airplane

15. _____ A timid, nervous student
    experiences extreme difficulty remem-
    bering stall entry procedures

16. _____ Because of several years of
    experience driving a car, a student
    forgets to maintain directional control
    with the rudder pedals during taxi
    operations

17. The most effective way to help an impatient student is to
    1. clearly establish the overall course goals.
    2. clearly identify the purpose and necessity for each intermediate step in the training course.
    3. increase the rate of introduction of new material.
    4. enliven the presentation of new material.

Match the words and phrases on the right with the appropriate defense mechanism on the left.

18. _____ Rationalization   A.   Daydreaming

19. _____ Flight   B.   Participation without interest

20. _____ Aggression   C.   Substitution of excuses for reasons

21. _____ Resignation   D.   Refusal to participate in class
                                               activity

Exercise 2B–
The Teaching Process

Match the words on the right with the appropriate step in the teaching process on the left.

1. _____ Preparation   A.   Summary—Progress Report

2. _____ Presentation   B.   Awareness—Perceptions—Insights—
                                               Habits
3. _____ Application
                                          C.   Objectives — Goals — Procedures —
4. _____ Review and evaluation   Facilities

                                          D.   Trial—Practice

5. _____ (True, False) A lesson plan may be thought of as a sequence of learning experiences designed to meet a particular instructional objective.

6. _____ (True, False) A carefully developed lesson plan can be used effectively for all students at the same stage of training.

7. _____ (True, False) Lesson plans should include a statement about minimum acceptable performance which reflects final goals rather than current experience levels.

8. Learning begins with awareness and continues until insights are formed into _____.

9. _____ (True, False) An evaluation of a student's performance must be based on the current lesson objectives, not the course completion standards.

10. For instructors to answer the question, "Did the student master the objectives to an acceptable performance level," they should refer to the part of the lesson plan labeled
    1. objectives.
    2. completion standards.
    3. contents.
    4. study assignments.

11. When an instructor relates the learning outcomes to career advancement, financial gain, or some other attraction, the student is being _____ to learn.

12. List four situations when the lecture method is appropriate in a training situation.
    1. _____
    2. _____
    3. _____
    4. _____

13. _____ (True, False) One disadvantage of the teaching lecture for flight training is that student feedback is indirect and difficult to interpret.

14. _____ (True, False) Because of the lack of student participation, the lecture method tends to foster passiveness.

15. List five types of questions that can be used in a guided discussion.
    1. _____
    2. _____
    3. _____
    4. _____
    5. _____

16. Final examinations are useful for determining final standing and for gaining information about the quality of the _____.

17. _____ (True, False) A critique is primarily the stage in the grading process that deals with the negative aspects of a performance.

18. _____ (True, False) The most important reason for a critique is to inform a student what must be done to improve future performance.

Exercise 2C–
Planning and Organizing

1. The length of a given lesson in the training syllabus should be established on the basis of
    1. an arbitrary time used for all lessons.
    2. units of learning.
    3. the size of the class.
    4. desired performance level.

2. In addition to objectives, contents, and study assignment, each flight training lesson should include
   1. building blocks.
   2. completion standards.
   3. teaching aids.
   4. time allotments.

3. _____ (True, False) A training syllabus contains an organized plan for instruction which the instructor must adhere to without deviation if the course objectives are to be attained.

4. _____ (True, False) A flight training lesson should include the instructional objective of the unit and the minimum acceptable performance.

Match the definition on the right with the corresponding term on the left.

5. _____ Building block

6. _____ Completion standard

7. _____ Syllabus

8. _____ Objective

9. _____ Stage of training

10. _____ Lesson plan

A. An abstract or digest of the course of training

B. The instructor's personal plan for adapting a single unit of training to a particular student

C. An organized presentation of material progressing from the simple to the complex

D. A statement of what is to be accomplished during a unit, stage, or course of training

E. The criteria used to determine whether an objective has been met

F. A grouping of common subjects, organized in a logical progression within a training syllabus

11. Audio-visual training is designed primarily to provide
    1. reinforcement.
    2. motivation.
    3. initial presentations.
    4. a basis for discussions.

12. _____ (True, False) Meaningful repetition is a necessary and desirable feature to incorporate into a course of training.

CHAPTER 3—INSTRUCTOR RESPONSIBILITIES
Exercise 3A—Authorized Instruction and Endorsements

1. A major quality of professionalism is
    1. an image of success.
    2. a code of ethics.
    3. to never make a mistake.
    4. to never admit being wrong.

Match the qualities of professionalism on the right with the descriptive statement on the left.

2. _____ Declining to fly an airplane with an inoperative fuel quantity indicator

3. _____ Reviewing the training syllabus prior to each dual flight

4. _____ Refusing to endorse a student's logbook for a flight test because the instructor has not personally trained the student

5. _____ Being able to adequately explain the difference between torque and P-factor

6. _____ Speaking to high school students about careers in aviation

7. _____ Demonstrating safe simulated emergency procedures

A. service is performed

B. training and preparation

C. study and research

D. intellectual requirement

E. good judgmental decisions

F. code of ethics

8. _____ (True, False) A flight instructor with an airplane rating only may not give dual instruction to an applicant for an instrument rating.

9. _____ (True, False) FAR Part 141 contains the requirements for certification of pilots.

10. A private pilot training under FAR Part 141 is required to have at least _____ hours total flight time, including _____ hours dual instruction and _____ hours solo flight time, of which _____ hours must be logged as cross country.

11. _____ (True, False) A flight instructor applicant applying for an airplane or instrument instructor rating must hold a commercial pilot certificate with an instrument rating.

12. _____ (True, False) A student pilot's certificate must be endorsed every 90 days for solo privileges.

13. An instructor must endorse a student pilot's logbook prior to each
    1. flight.
    2. solo flight.
    3. solo cross country.
    4. night flight.

14. _____ (True, False) The instructor is required to endorse the student's logbook each time dual instruction is given.

15. _____ (True, False) An endorsement is required by regulation for a flight review, regardless of the performance of the pilot being reviewed.

16. The requirement that flight instructors keep records for a period of at least three years applies to
    1. flight instruction exclusively.
    2. both flight and ground instruction.
    3. student pilot endorsements and test recommendations.
    4. any instruction provided under FAR Part 61.

17. _____ (True, False) Professional instructors should concentrate on giving excellent instruction and should not become involved in the management considerations of the operator who employs their services.

Exercise 3B–Regulations (FAR Part 61, Subpart G)

1. _____ (True, False) To be eligible for a flight instructor certificate, an applicant must hold a commercial or airline transport pilot certificate.

2. An applicant for a flight instructor certificate must present evidence showing satisfactory completion of a course of study that included student evaluation, _____, and _____.

3. The required flight instruction for a flight instructor applicant must be conducted by an instructor who has held an instructor certificate for at least the preceding _____ months and who has given at least _____ hours of instruction.

4. _____ (True, False) Each flight instructor is required to maintain personal records of all flight and ground instruction given for at least three years.

5. A flight instructor may not conduct more than _____ hours of flight instruction in any _____-hour period.

6. A flight instructor may not conduct flight instruction in any aircraft for which a category, _____, and (if appropriate) _____ rating is not held.

7. _____ (True, False) A flight instructor may endorse an applicant's student pilot certificate for a solo cross-country flight without having provided any instruction to the applicant as long as the student has received the required training from another instructor.

8. Required instruction for the issuance of a certificate or a category, or class rating in a multi-engine aircraft may not be given unless the instructor has _____ hours as pilot in command in the make and model airplane to be used.

9. _____ (True, False) An applicant for a flight instructor certificate who does not hold a commercial pilot certificate may receive a limited flight instructor certificate.

10. The required record of student endorsements may be kept in the flight instructor's _____.

11. _____ (True, False) A flight instructor must have the appropriate ground instructor ratings to conduct ground instruction.

12. _____ (True, False) The holder of a flight instructor certificate who applies for an additional rating on that certificate must have at least 15 hours of PIC time in the category and class of aircraft appropriate to the rating sought.

13. A flight instructor certificate may be renewed by attending an approved flight instructor refresher course which includes at least _____ hours of ground instruction, flight instruction, or both.

14. _____ (True, False) Flight instructor records must include the results of all tests for which students have been recommended.

CHAPTER 4—THE INSTRUMENT
FLIGHT INSTRUCTOR
Exercise 4A—
Basic Instrument Instruction

1. Gyroscopic instruments may be _____ or _____ powered.

2. _____ (True, False) The gyro instruments will provide accurate indications immediately after the engine is started.

3. _____ (True, False) The attitude indicator gives a direct indication of aircraft performance.

4. The attitude indicator normally erects at the rate of _____° per minute.

5. _____ (True, False) The attitude indicator may, during rapid acceleration, indicate a greater pitch-up altitude than actually exists.

6. _____ (True, False) Many attitude indicators are subject to operating limitations in both pitch and bank.

7. _____ (True, False) The attitude indicator is the only instrument that provides immediate and direct pitch and bank information.

8. For instrument flight, the heading indicator should be checked with the magnetic compass at least every
   1.   5 minutes.
   2.   10 minutes.
   3.   15 minutes.
   4.   20 minutes.

9. _____ (True, False) A heading indicator with a precession error of no more than three degrees in 15 minutes normally is considered acceptable for instrument flight.

10. _____ (True, False) Both the ball and turn-rate indicator of a turn coordinator are operated by gyros.

11. The miniature airplane of a turn coordinator deflected halfway to the reference indicates a turn at
    1.   1½° per second.
    2.   3° per second.
    3.   5° per second.
    4.   6° per second.

12. List the pitot-static system instruments.
    1. _____
    2. _____
    3. _____

13. Use of the alternate static pressure source may cause incorrect instrument readings. To determine this error, the _____ _____ _____ should be consulted.

14. Flight from a warm airmass to a cold airmass at a constant indicated altitude, using a constant altimeter setting, will result in an indication that is _____ (higher, lower) than the actual altitude.

15. _____ (True, False) The effects of nonstandard conditions can create errors of as much as 1,000 feet between indicated altitude and true altitude.

16. _____ (True, False) The vertical speed indicator provides instantaneous rate information.

17. The magnetic compass readings are accurate only during _____ _____ _____ , unaccelerated flight.

18. The inherent errors of the magnetic compass include variation, _____ , and _____ _____ .

19. When turning left from a northerly heading, the magnetic compass initially will indicate a turn to the _____ .

20. The three main elements of attitude instrument flying are scan, _____ and _____ .

21. If an airplane has a 20° bank angle during a turn, the roll-out lead when recovering to straight flight should be approximately _____ °.

22. The principal pitch control instrument for airspeed transitions is the _____ _____ .

23. _____ (True, False) Recoveries from unusual attitudes are not considered complete until the airplane has been returned to a stabilized level flight attitude.

Exercise 4B–Instrument Procedures Instruction

1. _____ (True, False) Before starting instrument procedures training, the instructor must be certain the student is proficient in holding pattern entry procedures.

2. _____ (True, False) If the static pressure system and the altimeter *have not been inspected* within the preceding 24-calendar months, the airplane may be operated under IFR provided the altimeter indicates within 75 feet of the departure airport field elevation when set to the local altimeter setting.

3. _____ (True, False) Takeoff minimums are provided for private aircraft primarily to insure that a flight will be able to return to the airport should serious mechanical difficulties occur during departures.

4. Departure training should include radar and nonradar departures, as well as

_____ _____ _____ .

5. In a no-wind condition, holding pattern legs are to be flown for one minute below an altitude of
   1. 10,000 feet.
   2. 12,500 feet.
   3. 14,000 feet.
   4. 18,000 feet.

Use the legend below to indicate the appropriate holding pattern entry procedures for each of the numbered aircraft. Assume the aircraft arrive at the VORTAC on the headings shown in the illustration.

A. Direct B. Teardrop C. Parallel

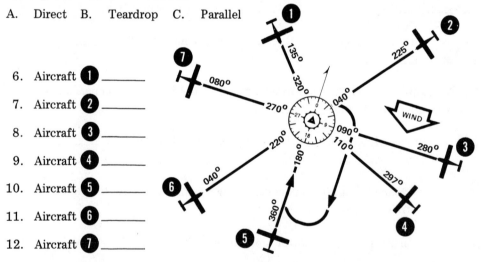

6. Aircraft ❶ _____

7. Aircraft ❷ _____

8. Aircraft ❸ _____

9. Aircraft ❹ _____

10. Aircraft ❺ _____

11. Aircraft ❻ _____

12. Aircraft ❼ _____

13. Assume an aircraft is in a standard holding pattern on the 270° radial of a VORTAC. If a magnetic heading of 080° is required on the inbound leg, what heading should be maintained on the outbound leg?
    1. 250°
    2. 260°
    3. 280°
    4. 290°

14. _____ (True, False) In every instance, direct descent to the initial approach altitude may be started as soon as approach clearance is received.

15. During a VOR approach with a procedure turn, the bearing selector should be set to the inbound course when
    1. crossing the VOR outbound.
    2. executing the procedure turn.
    3. intercepting the inbound course.
    4. at the VOR inbound.

16. The distance criteria for remaining within the circling approach protected area should not be based on the _____ _____ for the approach.

17. _____ (True, False) In VOR/DME approach procedures, an elapsed time required to fly to the MAP may not be provided because DME provides continuous fixes.

18. To intercept a DME arc at groundspeeds under 150 knots, a lead of _____ mile(s) is sufficient.

19. _____ (True, False) Unless the pilot is highly proficient with other airborne equipment, the FAA recommends the use of RMI equipment for DME arcs.

20. _____ (True, False) When the notation "VOR/DME" is included in the approach procedure identification of an approach chart, it indicates that DME equipment is mandatory for the approach.

21. The accuracy of an NDB approach is largely dependent on how accurately the _____ _____ is set.

22. _____ (True, False) A student should have a thorough understanding of ADF equipment and the information it presents before any attempt is made to fly an NDB approach.

23. A good method of introducing ILS procedures to a student is to ignore the glide slope and concentrate on _____ tracking.

24. It is imperative that the instructor emphasize the necessity for small _____ _____ on the localizer.

25. The instructor should point out the importance of having airspeed and altitude stabilized prior to _____ _____ interception.

26. _____ (True, False) ILS approach training is considered easier for the student than VOR or NDB approaches.

27. _____ (True, False) If a terminal radar facility is not available for ASR approaches, the instructor should simulate the approach instructions.

28. _____ (True, False) During an ASR approach, the controller automatically provides recommended altitudes on final approach.

29. _____ (True, False) During an ASR approach, the controller provides glide path information.

30. _____ (True, False) During a contact approach, it is necessary to remain clear of all clouds during the approach, just as if the flight were operating under VFR with a "special VFR" clearance.

31. An IFR alternate airport must be selected if the forecast ceiling at the destination within one hour of the ETA is less than _____ feet above the airport elevation and the forecast visibility is less than _____ mile(s).

32. _____ (True, False) When airborne, those parts of an ATC clearance containing altitude assignments or vectors should be read back to the controller for mutual verification.

33. During cross-country flights the instructor should request revised _____ for reporting points and fixes.

34. _____ (True, False) Pilots must report over compulsory reporting points in both radar and nonradar environments.

35. A pilot can maintain instrument currency through recent experience or an
_____ _____ _____ .

36. An instrument competency check consists of
    1. oral and written tests.
    2. written and flight tests.
    3. oral discussion and flight test.
    4. oral discussion and review flight.

CHAPTER 5—THE MULTI-ENGINE FLIGHT INSTRUCTOR

Exercise 5A—Multi-Engine Airplane Instruction

1. An important aspect of the transition to a multi-engine airplane is complete familiarity with the _____ of the airplane.

2. The first multi-engine flight lesson should consist of basic airplane _____.

3. Correct taxi procedures include the proper use of _____ power.

4. To properly perform takeoffs and climbs, the multi-engine student must have a thorough understanding of all _____ speeds.

5. _____ (True, False) A beginning student usually has little problem maintaining airspeed control during the takeoff and initial climb.

6. During takeoff, the instructor must be alert for any uneven _____ application.

7. After takeoff the multi-engine airplane should be accelerated as rapidly as possible to the best _____-_____-_____ airspeed.

8. _____ (True, False) When the pilot's operating handbook outlines a short-field takeoff using a liftoff speed below V_{MC}, that speed should be used.

9. _____ (True, False) Common student errors during multi-engine maximum performance takeoff training are improper airspeed and power control.

10. Prior to beginning practice in maximum performance takeoffs, the student should display the highest degree of proficiency in
    1. power control.
    2. airspeed control.
    3. systems management.
    4. maneuvering ability.

11. The major difference between takeoffs in single-engine and multi-engine aircraft is the
    1. basic takeoff technique.
    2. plan of action that is necessary in case of engine failure.
    3. point of gear retraction.
    4. different airspeeds and flap settings.

12. _____ (True, False) Use of consistent traffic pattern procedures and profiles increases safety and helps establish good operating practices.

13. The disadvantages of a power-off approach in a twin are _____ approach attitudes and _____ rates of descent.

14. The correct drag "cleanup" sequence for a specific airplane during a go-around can be found in the _____ _____ _____ .

15. In the event of a go-around, the aircraft should be accelerated to the best _____-_____-_____ speed.

16. Maneuvering at critically slow airspeed enables the student to learn the effect of the _____ and _____ _____ on airplane performance.

17. Two common errors that are evident during slow flight practice are failure to control _____ and _____ .

18. _____ (True, False) While maneuvering at critically slow airspeed, the instructor must emphasize that airplane attitude will change as landing gear and flaps are extended.

19. During recovery from an imminent stall in the landing configuration, the student may be asked to recover to a specific _____ and _____ .

20. If the student does not apply equal power on both engines during imminent stall recovery, serious _____ _____ difficulties may arise.

21. Steep power turns help the student develop a feel for the control _____ of the airplane.

22. During steep power turns, the student frequently has difficulty controlling _____ .

23. _____ (True, False) The student should be taught to trim the airplane to fly "hands off" in 50° steep bank turns.

24. Engine out procedures should be introduced when the student has become completely familiar with the basic _____ _____ of the airplane.

25. _____ (True, False) Single-engine maneuvers initially should be done with one propeller windmilling.

26. _____ (True, False) An airspeed higher than single-engine best rate-of-climb will increase climb performance.

27. _____ (True, False) The student should memorize the in-flight engine shutdown checklist.

28. The instructor must emphasize the importance of using the correct procedure for identifying the _____ _____ .

29. During an engine inoperative loss of directional control demonstration, the instructor must not allow the student to establish an abnormally high _____ _____ .

Exercise 5B–Aerodynamics of Multi-Engine Flight

1. _____ (True, False) Thrust in conventional multi-engine airplanes is not directed along the airplane's center line.

2. _____ (True, False) Induced airflow is an important advantage of wing-mounted engines.

3. _____ (True, False) The multi-engine service ceiling is the maximum density altitude at which the best angle-of-climb airspeed will produce a 100 f.p.m. climb.

4. _____ (True, False) The best angle-of-climb speed increases with altitude.

5. _____ (True, False) When one engine fails, the loss of climb performance is only 50 percent.

6. The factor creating the greatest amount of drag during an engine-out situation usually is
   1. fully extended flaps.
   2. extended landing gear.
   3. a windmilling propeller.
   4. unique to the specific type of multi-engine airplane.

7. The term that refers to the engine whose failure most adversely affects the performance or handling characteristics of the airplane is _____ _____.

8. In conventional twin-engine airplanes, the most adverse affects on the handling characteristics are caused by failure of the _____ engine.

9. The minimum airspeed at which airplane control can be maintained with the critical engine inoperative and the other engine producing takeoff power is known as _____.

10. Banking the aircraft toward the operative engine will produce a
    1. lift vector on the low wing which relieves the rudder forces.
    2. lift vector on the high wing which relieves the rudder forces.
    3. drag vector on the low wing which relieves aileron pressure.
    4. drag vector on the high wing which relieves aileron pressure.

11. _____ (True, False) V_{MC} is affected by loading.

12. What effect, if any, will loading the airplane with the CG at the aft limit have on actual V_{MC} as compared to a CG located at the center of the range?
    1. No effect
    2. Reduces V_{MC}
    3. Increases V_{MC}
    4. Effects V_{MC} only if loaded above the maximum gross weight

13. If an engine-out climb is not possible, the slowest rate of descent can be attained by using what airspeed?
    1. V_{MC}
    2. V_Y
    3. V_{XSE}
    4. V_{YSE}

14. The maximum density altitude at which V_{YSE} produces a 50 f.p.m. rate of climb is called the
    1. service ceiling.
    2. single-engine service ceiling.
    3. absolute ceiling.
    4. single-engine absolute ceiling.

15. An engine-out climb is not possible if
    1. lift exceeds drag.
    2. drag exceeds thrust.
    3. thrust exceeds lift.
    4. drag and lift are equal.

16. _____ (True, False) If a spin is entered in a multi-engine airplane as a result of improper stall recovery or engine-out loss of directional control demonstration, recovery is highly unlikely and, in some cases, impossible.

PILOT BRIEFINGS

INTRODUCTION

Pilot briefings are a series of essay-type questions designed to provoke further study and discussion of training concepts. Each briefing is assigned at a strategic point in the training syllabus. Prior to the briefing session, the applicant should complete the appropriate exercise by writing the answer to each question on a separate sheet of paper. During the briefing, the applicant and instructor will review each question and answer, then discuss related information. Answers to most of the pilot briefing questions are found in the textbook, while others may be found in the pilot's operating handbook or other appropriate publications.

FUNDAMENTALS OF INSTRUCTION
THE LEARNING PROCESS

1. Can a "beginning" student profit most from an evaluation of performance according to a standard or from an individual critique? Why?

2. Briefly describe the law of readiness as it pertains to a training situation.

3. List some of the circumstances which may undermine the instructor's ability to stimulate the student's readiness to learn.

4. Briefly explain the law of exercise as a part of the general laws of learning.

5. What teaching method is based on the law of exercise?

6. The law of effect is based on the emotional reaction of the learner. Briefly describe the principles of this law.

7. What experiences tend to impede further learning in a specific area?

8. What does the law of primacy mean to an instructor?

9. Generally, how can an instructor insure that flight and ground presentations give students the advantages of the law of intensity?

10. What theories of forgetting are most applicable to flight training? Briefly explain each.

11. Most theories of forgetting imply that forgotten information is not totally lost, but simply is not immediately available for total recall because of disuse, interference, or repression. List several teaching methods the instructor can use to enhance learning and minimize forgetting.

12. Is a serious problem indicated if continued practice does not produce further obvious improvement for the beginning student?

13. Why does the student's capacity for understanding increase as skill is acquired through practice in a given area?

14. Does the occurrence of a learning plateau mean that learning has ceased?

15. Occasionally, a learning plateau is reached and further student improvement seems unlikely. What may this signify?

16. What can be done to ward off discouragement in a student *before* a leveling off of progress occurs?

17. What term best describes the phenomenon that causes the learning of one procedure to hinder the learning of another?

18. How can instructors use a knowledge of positive and negative transfer to train students more efficiently?

19. What is the key to teaching students to transfer what they already know to new situations?

20. List several obstacles to student learning which are common in flight instruction.

21. What is the basic reason for the building-block technique of instruction?

22. List, in sequence, the four general levels of learning.

23. What two conditions must be present before a student can be expected to *use* a skill that has been learned?

24. Briefly state three characteristics of instructional objectives.

25. List six factors that may modify or alter the perceptions that form the basis of all learning.

26. What is the relationship between student goals and instructional objectives?

27. What guidance does the typical motor skills learning curve provide for the flight instructor?

28. What are the consequences of allowing a student to continue practicing a bad habit?

29. List four human relations concepts that can help to improve the skill of a flight instructor.

30. Lack of common experience is a barrier to communications. What can the flight instructor do to overcome this barrier?

31. Does the idea that people can be motivated to learn by making it easy for them have any basis in fact?

32. Failure to perform may be justified by rationalization. How can rationalization be recognized?

THE TEACHING PROCESS

1. Why is it important for an instructor to conduct a private interview with a student prior to beginning flight instruction?

2. What value is there in letting students discover a certain percentage of flying skills on their own during directed practice?

3. What general category of student is least likely to benefit from the "discovery" technique of instruction?

4. What are the two primary guides which provide the instructor with a clear, step-by-step plan of instruction?

5. Why must an instructor keep students continuously informed of their progress?

6. Why should the instructor demonstrate a maneuver in which the student has experienced some difficulty?

7. Why is it essential that a student's mistakes be corrected immediately?

8. How should a flight instructor deal with incipient airsickness?

9. What are the cues an instructor should watch for to determine the onset of student fatigue?

10. How might an instructor best handle the impatient student who fails to understand the need for preliminary training and seeks only the ultimate objective?

11. Generally, how can an instructor deal with personal student interests, fears, enthusiasms, and problems over which the student has no control?

12. Why are speeches to inform, persuade, or entertain of little value to a flight instructor?

13. What is the preferred delivery method for a teaching lecture?

14. What two benefits are derived by a student in the process of explaining a maneuver as the instructor performs it?

15. What important student learning concept is being taught by the "student tells—student does" step of instruction?

16. What important limitation is imposed on the use of visual materials for teaching purposes?

17. What is the main learning advantage of linear programmed instructional material?

18. What factors should be considered when setting lesson performance goals for the individual student?

19. What is the best way to continually motivate students?

20. What can be done by the instructor to foster the proper instructor/student relationship?

21. Arranging subject matter chronologically is one way to develop a lesson plan. Name three other ways to organize material suitable for flight training.

22. Under what circumstances is the lecture method a suitable instructional approach to flight training?

23. Describe the difference between an overhead and a relay question.

24. What types of instructional aids are compatible with the learning objectives of a pilot training course?

25. Is a critique a part of the grading process or a part of the learning process? Explain.

26. What characteristics do effective critiques possess?

27. What are the ground rules which must be observed for a critique to accomplish its intended purpose?

28. Through trial and practice a student is guided into development of insights and formation of habits. Which one of the four basic steps in the teaching process does this activity represent?

PLANNING AND ORGANIZING

1. Rearrange the following training procedures and maneuvers in the logical building-block order of presentation that would be used for a student pilot.
   Tracking a straight line between two points
   Turns, climbs, and descents
   Straight-and-level flight
   Normal and crosswind takeoffs and landings
   Maneuvering at critically slow airspeed

2. What purposes should a lesson introduction accomplish?

3. A flight syllabus which uses previously learned maneuvers as a basis for subsequent maneuvers takes maximum advantage of what kind of transfer?

4. How are the relationships between main points usually shown when developing a lesson?

5. What three steps should be included in the introduction to a guided discussion?

6. Why must objectives be stated in terms of observable, definable, and measurable behavior?

7. Why is it impossible to completely standardize instruction?

8. In what two areas does a critique significantly differ from an evaluation?

9. What are the *overall* objectives an instructor should strive to attain while instructing student pilots?

10. Briefly explain the reasoning behind teaching in graduated blocks of learning.

11. Describe how learning blocks of instruction should be divided to produce lesson plans.

12. What are the criteria for determining the useful size of a block of learning?

13. Explain the format and importance of a syllabus to a flight student. Should it ever be altered?

14. What extra responsibilities are placed on an instructor when a syllabus is altered?

15. Providing ground instruction to one individual at a time has several advantages. Explain why scheduling is an advantage of such instruction.

16. Why does individual ground instruction enhance student progress?

17. What is probably the greatest benefit of the formal classroom environment for ground training?

18. One disadvantage of classroom instruction is the requirement for enough students enrolled to warrant starting the course. What effect does this have on students already enrolled in the training program?

19. In which instructional situation is oral quizzing most effective?

20. What action should be taken by the instructor if a student is having a problem with a certain phase of training?

FLIGHT INSTRUCTOR ORAL QUESTIONS
AERODYNAMICS OF FLIGHT

1. What are the four equal and opposing forces that act upon an airplane in unaccelerated, straight-and-level flight?

2. What aerodynamic factor must be changed to compensate for weight or G-loading?

3. Name four factors which influence total lift.

4. What power relationship determines the maximum rate-of-climb speed?

5. Generally, how are the three aircraft axes identified and what are they named?

6. Which aircraft movement is associated with each of the axes?

7. Briefly explain the production of lift by an airfoil.

8. Why are most wings designed so the root will stall before the wingtip?

9. Name two methods used by manufacturers to cause the wing root to stall first.

10. Briefly define dynamic stability.

11. What causes adverse yaw?

12. What have manufacturers done to minimize adverse yaw?

13. How does a level turn affect the load factor of an aircraft?

14. What is the significance of maneuvering speed?

15. Describe the maximum angle-of-climb airspeed.

16. Describe the direction of movement of relative wind.

17. Why does the stall speed often differ from the airspeed indicated at the low speed end of the white arc on the airspeed indicator?

18. How does moving the CG forward or aft affect stability?

19. What is one of the most serious effects of ground effect on takeoff?

20. What is the advantage of a controllable pitch propeller?

21. Why does induced drag decrease as angle of attack decreases?

22. What causes total drag to be at a minimum at one specific airspeed?

23. Why is parasite drag greatest at high airspeeds?

BASIC MANEUVERS

1. What precautions should be taken when taxiing a tricycle-gear airplane in strong winds?

2. How does this taxi technique change if the aircraft is tailwheel equipped?

3. What precautions are necessary when landing or taxiing a light airplane behind a large airplane that is being run up on the ground?

4. What common student error causes the nose of an airplane to inscribe an arc above or below the horizon during a turn entry or recovery? How can this error be corrected?

5. What corrective action is needed if a steep turn develops into a power-on spiral?

6. When the control forces are released during a shallow bank, why does the airplane tend to return to level flight?

7. Why does bank increase if aileron control pressure is relaxed during a steep bank turn?

8. Can a steep bank, power-on spiral be corrected by use of back pressure alone? Why?

9. Besides engine and carburetor heating, what factors might require the use of power during a prolonged descent?

10. What variable flight conditions can affect the stalling speed of an airplane?

11. What factors determine the bank angles needed during turns around a point?

12. Describe the proper recovery from a nose-low, turning, unusual flight attitude.

13. What vital responsibility frequently lapses while the pilot is concentrating on ground reference maneuvers?

14. How is the scanning technique that is used during the day modified for effective scanning at night?

15. Why might full engine power during flight at critically slow airspeed be insufficient to produce a climb when the airplane can climb readily at a higher airspeed? What term frequently is used to describe this condition?

16. How can a pilot operating on the back side of the power curve establish a climb?

17. Why is the ability to recognize an imminent stall important for safety of flight?

18. Why are ground reference maneuvers practiced and what is their chief use during normal flight?

19. What visual references are commonly used to maintain proper pitch and bank attitude in straight-and-level flight?

20. What common faults cause the student to maintain straight flight with one wing low?

21. Explain why rudder forces must be used when entering and recovering from maneuvers which require aileron displacement.

22. When practicing S-turns across a road that is perpendicular to the wind, at what point should the bank angle be greatest?

23. What technique is used to correct for wind during turns around a point?

SHORT-FIELD AND SOFT-FIELD TAKEOFFS AND LANDINGS

1. What general procedures are used to make a takeoff from a hard-surfaced short field without obstacles?

2. How are these procedures modified if obstacles are present in the departure path?

3. Describe the technique normally used to begin the takeoff roll from a rough or soft field.

4. Why is it inadvisable to maintain the pitch attitude used for a soft-field takeoff during the initial climb?

5. Besides runway surface conditions, what factors can greatly affect runway length requirements?

6. What is the significant advantage achieved by using best angle-of-climb airspeed during a takeoff over an obstacle?

7. What allowances should be made for gusty wind conditions during a short-field approach?

8. What techniques are used for a short-field landing over an obstacle?

9. What effect does the use of power have on the stabilized short-field approach?

10. When landing at a small airport without a control tower, FSS, or UNICOM, what indicators can be used to determine the wind direction?

11. Describe a normal landing.

12. How will a moderate slip affect the indicated airspeed?

13. What are the basic objectives in practicing simulated emergency landings?

14. What improper control usage is most likely to upset the stabilized approach over an obstacle?

15. How does the soft-field landing technique vary from the short-field procedure?

16. What control technique should be used to attain the fastest acceleration on a short hard-surfaced runway?

17. How do the takeoff procedures for soft or rough fields differ from those used on short, hard-surfaced runways?

18. Will the procedures for a takeoff from a soft field without obstacles vary from those used for a soft-field takeoff with obstacles?

19. After touchdown, what technique can be used to shorten the landing roll?

20. Is the short-field landing distance specified in performance chart figures based on maximum braking?

ADVANCED MANEUVERS

1. Describe a practical application of steep spirals.

2. Describe the entry procedure for a steep spiral.

3. In general terms, describe the steep power turn.

4. What are the common causes of altitude variation during a steep turn?

5. What rudder control forces are required during a steep turn to the left? How do they differ during a steep turn to the right?

6. What instruments are used to monitor altitude during a steep turn?

7. What benefits are derived from practice of chandelles?

8. Describe the aircraft attitude changes which occur during execution of a chandelle.

9. What elevator pressure changes occur during the chandelle?

10. What common error causes the bank angle to decrease as the pitch attitude increases during a chandelle?

11. What results can be expected if the angle of bank is too steep during a chandelle entry?

12. What affect will a strong crosswind have on the lazy eight pattern?

13. At what point is the pitch attitude highest during performance of a lazy eight?

14. How will the lazy eight pattern change if the selected reference point is too close to the aircraft?

15. What benefits are derived from performance of a lazy eight?

16. How does aft CG loading affect airplane spin characteristics?

17. What procedures are used to ensure adequate engine heating during an intentional spin?

18. What reaction can be expected if full rudder is applied before the airplane enters a full stall during spin entry?

19. Are the characteristics of spins to the left identical to those of spins to the right? Why?

20. Describe the general spin recovery technique.

21. Should power be used during the time the airplane is spinning?

22. What reaction can be expected if the rudder and elevator control forces are relaxed during a spin?

23. Why is it important to monitor the airspeed closely during spin recovery?

EMERGENCY OPERATIONS

1. What corrective action is required if the ammeter shows a continuous discharge in flight?

2. What common in-flight problem is indicated by a gradual loss of r.p.m. and eventual engine roughness? What corrective action is required for this condition?

3. What procedure should be followed to alleviate fouled spark plugs?

4. What usually is signified by a low oil pressure indication with no significant rise in oil temperature?

5. What does a total loss of oil pressure and rising oil temperature indicate?

6. If detonation or preignition is detected, what corrective procedures should be followed?

7. If a pilot who is not instrument rated inadvertently enters IFR weather, what general technique will enhance a safe return to VFR conditions?

8. Outline some general guidelines for using the ELT during an emergency.

9. What steps should be taken to prevent fuel contamination?

10. Why does an encounter with wake turbulence usually create an emergency situation for small aircraft?

11. What can be done to avoid the hazard of jet blast?

12. What procedure should be used for making a precautionary landing with engine power?

13. What general procedure should be used for making an emergency landing without engine power?

14. What are the recommended procedures for obtaining emergency assistance?

15. What procedure should be followed if the landing gear will not retract after takeoff?

16. Describe the landing gear emergency extension procedure.

17. What is the proper procedure for a landing with a flat main landing gear tire?

18. What is the proper procedure for landing with a flat nose gear tire?

GENERAL SUBJECTS

1. During oral quizzing, what types of questions should be avoided?

2. What action should be taken by an instructor if there is a suspicion that a student does not understand

a maneuver, even though performance of the maneuver is correct?

3. Explain the difference between a lesson in a training syllabus and a lesson plan prepared by a flight instructor.

4. When are slumps or plateaus in the rate of learning most likely to occur?

5. Why should the student practice maneuvering at critically slow airspeed?

6. How is a forward slip normally used?

7. Explain P-factor.

8. How does the instructor know when a student is prepared to solo?

9. What causes ground effect?

10. What are some of the most common items which produce student fatigue?

11. What are the four fundamentals of flight, and why must the student understand them thoroughly?

12. Are control pressures or control movements more desirable?

13. Why are medium turns taught prior to steep or shallow turns?

14. What determines the radius of turn?

15. What are the most common errors when taxiing an airplane?

16. Define best rate-of-climb speed.

17. Define best angle-of-climb speed.

18. Why should the lesson plan be discussed with the student prior to takeoff?

INSTRUMENT FLIGHT INSTRUCTOR ORAL QUESTIONS

ATTITUDE INSTRUMENT FLYING

1. A consistent loss of altitude during steep turn entries indicates a lack of understanding of what basic requirement?

2. What are the three conditions that must be met before a pilot may descend below the MDA or DH during an instrument approach?

3. Assume a flight at 3,500 feet will cross an intersection in six minutes. If the pilot is issued an IFR clearance to cross the intersection at 8,000 feet, what average rate of climb must be used to comply with the clearance?

4. When navigating to an intersection identified by an airway and a VOR radial, why should the OBS be set to the published VOR radial, rather than to its inbound bearing?

5. During a constant descent maneuver, such as an ILS approach, what control is used to correct glide path deviations?

6. What is indicated if the attitude indicator shows that the aircraft is sharply nose up, but the airspeed is not decreasing and the altitude is not increasing?

7. What is the initial objective during recovery from a nose-high unusual flight attitude?

8. Describe the proper recovery from a nose-high, turning, unusual flight attitude.

9. If the attitude indicator is inoperative, what instruments are used for pitch information?

10. What is the procedure for determining the approximate angle of bank required for a standard-rate turn?

11. Outline the proper procedure for leveling off from a climb made by instrument references.

12. What is the significance of the term *cruise* preceding an altitude assignment from ATC?

13. During a timed turn to the right, the turn needle indicates the correct rate of turn, but the ball is to the left of center. What is the appropriate correction procedure?

14. If a flight is cleared to a fix short of the destination airport and is not given an expect further clearance time, when may further clearance normally be expected? What action is required if further clearance is not received?

15. Compare the sensitivity of the CDI when used for localizer functions and for VOR navigation.

16. Under what circumstances is reverse sensing of the CDI unavoidable?

17. When operating at the MOCA, navigational signal coverage is insured for what distance from a VOR?

18. List the recommended techniques for trimming the aircraft during instrument flight.

19. How does proper trim technique result in more precise, smoother airplane control during attitude instrument flying?

20. Explain how changes in altitude, power, or configuration may require a trim adjustment.

21. What is the most likely cause if a student "chases" the instrument indications and lacks precise airplane control?

22. How are pitch, bank, and power used to maintain straight-and-level unaccelerated flight?

23. During an ILS final approach, what controls are used to maintain the glide slope and airspeed?

INSTRUMENT CHARTS

1. What does the abbreviation MAA represent? Why is it used on some airways?

2. How is an airport with a published instrument approach procedure depicted on a low altitude chart?

3. Why are preferred routes used?

4. Discuss the parameters of a positive control area and the distinctive equipment requirements for flight in this airspace.

5. What are SIDs and STARs?

6. Compare DH and MDA.

7. What is the meaning of the notation NoPT on approach charts?

8. Discuss the initial approach fix (IAF) and final approach fix (FAF).

9. Why should a procedure turn be executed on the maneuvering side of the approach course, as depicted on the approach chart?

10. What is an area navigation (RNAV) *waypoint*?

11. Why are holding patterns depicted at some fixes on instrument charts?

12. Discuss the use of runway visual range (RVR) and prevailing visibility for landing minimums.

13. Why are geographical coordinates given below the VOR facility boxes on Jeppesen high altitude enroute charts?

14. How are remote communications outlets designated on charts?

15. Explain the significance of minimum enroute altitude (MEA) and minimum obstruction clearance altitude (MOCA).

16. Assume a pilot is cruising at an MEA of 7,000 and is required to report an arrival at a specific intersection. What action is indicated by a small MRA symbol and the designation "8,500" near that intersection?

17. Discuss the significance of a minimum crossing altitude (MCA) on an enroute chart.

18. What is the standard width of an airway? Are there any exceptions to this general rule?

19. Are NDB magnetic bearings and VOR radials determined as courses to or from the station?

20. Discuss special use airspace as shown on charts.

21. If Special VFR is not authorized at an airport, how is this indicated on NOS and Jeppesen charts?

22. Describe the three types of VOR navaid facilities.

23. How are MEAs and MOCAs determined for uncontrolled airspace?

24. What symbol is used to designate a change in the MEA or MOCA along a route?

INSTRUMENT APPROACH PROCEDURES

1. Under what circumstances must an alternate airport be listed on an IFR flight plan?

2. What weather minimums must be present at an airport before it can qualify as an alternate in a flight plan?

3. Under what circumstances may an airport without a published instrument approach procedure be used as an alternate?

4. How can pilots signify that they do not wish to be issued a STAR?

5. In terms of a DME arc transition, what is the purpose of a lead-in radial?

6. How does the pilot determine the missed approach point (MAP) during a nonprecision approach? How is it determined during a precision approach?

7. What significant advantages do ASR and PAR approaches have over other types of instrument approaches?

8. What are feeder routes? Why are they used?

9. Why are the weather minimums required for nonprecision approaches higher than those for precision approaches?

10. List the appropriate length for the inbound leg of a holding pattern according to altitude.

11. What requirements must be met before a contact approach clearance can be issued?

12. What are the requirements for a visual approach?

13. Under what conditions is a circling approach used?

14. What are the three holding pattern entry procedures? What determines which one is used for a specific entry?

15. Why must the holding clearance include an EFC time?

16. What are the basic ground components of a Category I ILS installation?

17. What is the relative CDI sensitivity when flying an ILS approach as opposed to a VOR approach?

18. Is the decision height always reached at the middle marker on an ILS approach?

19. During an ILS approach, what aircraft control is used to maintain airspeed? What is used to hold the aircraft on glide slope?

20. When a DME arc is used to transition for an instrument approach, is a procedure turn authorized?

21. Describe briefly the difference between the runway alignment indicator light system (RAIL) and the runway end identifier light system (REIL).

22. Where is the visual descent point (VDP) located on the final approach course of a straight-in nonprecision approach procedure?

23. What is the difference between an LOC and an LDA approach?

24. What substitution can be made for the outer marker (OM) of an ILS?

25. What is the meaning of the "runway environment"?

IFR EMERGENCY PROCEDURES

1. What facilities can be used by ATC to give instructions when communications radios are inoperative?

2. When communications are lost in IFR conditions and VFR conditions are encountered at a later time, what procedure should be followed?

3. If radio contact with the controlling agency is lost while flying IFR, how should contact be attempted?

4. What are the three main items to check while troubleshooting a radio failure in flight?

5. Which instruments may be used to replace the functions of an inoperative heading indicator?

6. What instruments will provide pitch information in the event of attitude indicator failure?

7. When is the correct time to begin the approach if radio communications fail while holding at a final approach fix?

8. Assume the student is using both navigation radios to hold at an intersection and DME is not available. How can the failure of one navigation radio be simulated and what procedure should be followed by the student?

9. Can a pilot execute a precision ILS approach with an inoperative marker beacon receiver?

10. Assume the student is making a simulated ILS approach in a radar environment. How can the instructor simulate a navigation radio failure and teach the student to transition to a radar approach?

11. If an enroute radio communications failure occurs, which route should be flown?

12. The alternator circuit breaker has "tripped" and, after waiting for it to cool, the pilot resets it. If the circuit breaker immediately trips again, what action should be taken?

13. When making an ILS approach, the runway lights are in view at the DH. However, as the pilot continues to descend, the lights are obscured partially and then disappear. What is the correct procedure?

14. What is the significance of the minimum safe altitude (MSA) designations on approach charts?

15. What types of weather encountered while on an IFR clearance must be reported to ATC?

16. What comprises total electrical failure? How can it be simulated?

17. What instruments become unusable if ice is present in the pitot tube? Can this be simulated in flight?

18. What is the lost communications procedure for transponder use?

19. What is the purpose of the two-way radio communications failure procedures in FAR Part 91.185?

20. If a pilot holding over an LOM experiences an ADF failure, how can the holding pattern be continued?

GENERAL SUBJECTS

1. Why should an instructor have a lesson plan for all phases of instrument flight instruction?

2. Why is the time allocation within the lesson plan important?

3. How can a lesson plan be revised to meet the needs of a student who is having difficulty with a certain phase of the lesson being taught?

4. How could a lesson plan be revised for a student who shows an exceptionally rapid grasp of the material presented?

5. How can student performance be evaluated on each lesson?

6. What method of evaluation should the instructor use following the completion of a flight lesson?

7. What techniques of evaluation would be most helpful to a student who is having difficulty with the course?

8. What techniques of evaluation can be used to maintain motivation in an exceptionally good student?

9. What standards should be used when evaluating student competency for an instrument rating practical test recommendation?

10. Outline the material to be covered during the preflight briefing for a flight lesson introducing ILS and localizer approaches.

11. Prepare a proper logbook entry for a commercial pilot flight test recommendation.

12. What should be contained in a log-book entry for a completed instrument competency check?

13. Outline some of the basic responsibilities of an instrument flight instructor regarding his students.

14. What level of performance must be demonstrated before a recommendation for an instrument rating is issued?

15. Explain the functions of the basic components of the ILS.

16. Present the elements of a postflight discussion and evaluation of student performance.

17. How does the use of the preflight briefing and postflight critique enhance the learning process?

18. What can be done in preflight and postflight instruction periods to stimulate and motivate better student performance?

19. What prebriefing preparation by the instructor can make the preflight and postflight instruction periods most effective?

20. What facilities should be readily available during the preflight and postflight instruction sessions?

MULTI-ENGINE FLIGHT INSTRUCTOR ORAL QUESTIONS

ANALYSIS OF MANEUVERS

1. What airspeed should be used for lift-off and climb when performing a short-field takeoff? Why?

2. What power setting is recommended during the execution of steep power turns?

3. Explain the procedure for setting up and executing imminent stalls in the gear-down and full-flaps, as well as gear-up and flaps-up, configurations.

4. Explain the procedure for maneuvering during slow flight.

5. In addition to a propeller synchronizer, what is another effective method of synchronizing the propeller r.p.m.?

6. After takeoff, what is the minimum altitude that must be attained before the power is reduced to climb thrust? Why?

7. List the following airspeeds and the corresponding airspeed indicator color codes, where applicable.
   1. Stalling speed with gear down, flaps down
   2. Stalling speed with gear down, flaps up
   3. Best angle-of-climb airspeed
   4. Best rate-of-climb airspeed
   5. Enroute climb airspeed
   6. Normal approach speed
   7. Maximum flap extension speed
   8. Maximum landing gear operating speed
   9. Maximum landing gear extended speed
   10. Design maneuvering speed
   11. Maximum structural cruising speed
   12. Never-exceed speed
   13. Minimum safe single-engine speed
   14. Best single-engine rate-of-climb airspeed
   15. Minimum single-engine control speed
   16. Maximum full flap extended speed

8. How does weight affect the stalling speed of the aircraft?

9. What gear/flap configurations and conditions are used to determine a published single-engine minimum control speed?

10. What is the proper recovery method if the pilot inadvertently allows the aircraft to decelerate below the minimum control speed during single-engine operations?

11. Explain the correct procedure for performing a short-field landing over an obstacle.

12. Assume the left engine is shut down, the propeller feathered, and the airplane trimmed for hands-off flight. What control problems may occur during landing if the trim setting is not changed?

13. Explain the correct procedure for drag cleanup during the recovery from a gear-down and flaps-down imminent stall.

14. If a student experiences difficulty in establishing and maintaining a stabilized landing approach, what technique can be used to correct the problem?

15. What student problem is indicated by the consistent use of a climb speed in excess of the best rate of climb immediately after takeoff? How can this error be corrected?

MULTI-ENGINE PERFORMANCE AND AERODYNAMICS

Performance and weight and balance questions in this Pilot Briefing are to be answered using the performance charts and graphs for the training aircraft.

1. Determine the total takeoff distance required to clear a 50-foot obstacle under the listed conditions.

   Temperature .... 24°C
   Airport elevation .... 2,700 ft.
   Altimeter setting 29.78
   Weight .... Maximum takeoff
   Surface wind .... Calm

2. What is the accelerate-stop distance if an engine failure occurs at the takeoff decision speed under the listed conditions?

   Temperature .... 18°C
   Pressure altitude .... 5,000 ft.
   Weight .... Maximum takeoff
   Headwind .... 5 kts.

3. Determine the weight and center of gravity location using the following data.

Pilot and front passenger.. 365 lbs.
Third and fourth seat
   passengers........... 220 lbs.
Fuel—main tanks .... Full
Baggage* .... 165 lbs.

*Assume that the most forward baggage compartment is loaded to its capacity first, followed by loadings in the next most forward compartment, etc.

4. Which engine on a conventional twin-engine airplane has the greatest effect on left-turning tendencies during a two-engine climb? Explain why this occurs.

5. When planning a takeoff in a twin-engine airplane, what performance factors will aid the pilot in determining the correct action to take if an engine fails?

6. Explain why the use of differential power on a conventional twin causes the airplane to turn toward the engine producing the least power.

7. Assume that a 500 f.p.m. descent rate is desired from cruise altitude to the traffic pattern altitude. How many miles are required for the descent, using the listed conditions?

   Cruise altitude........ 13,500 ft.
   Traffic pattern altitude .. 1,250 ft.
   Descent groundspeed..... 195 kts.

8. Why should a pilot add an extra margin of safety to the performance values determined from charts or graphs?

9. Why does lower-than-standard atmospheric pressure result in a loss of performance for nonturbocharged engines?

10. What is the most likely cause of a manifold pressure overboost when operating turbocharged engines with automatic wastegates? How can an overboost be prevented?

11. In a turbocharged engine with the wastegate fully closed, what effect does an engine r.p.m. change have on the manifold pressure?

12. Define induced and parasite drag. In what general airplane ranges does each type of drag predominate?

13. Explain why the use of the alternate induction air system on a fuel injected engine results in a small loss of engine power.

14. Generally, what airplane component or components limit the landing gear extension speed?

15. Define multi-engine service ceiling and explain how this differs from the single-engine service ceiling.

ENGINE-OUT PERFORMANCE AND AERODYNAMICS

1. Assume an engine has been shut down and its propeller feathered. If the cylinder head temperature on the operating engine is approaching the red line, what procedures can the pilot use to alleviate this condition?

2. If all oil pressure is lost on one engine, what probable events are likely to follow?

3. If the left engine fails on a conventional twin, which rudder must be applied to counteract yaw?

4. In a conventional twin, why does a five-degree bank toward the operative engine decrease the amount of rudder deflection required to counteract yaw?

5. Explain why the *actual* V_{MC} speed of a nonturbocharged conventional twin-engine airplane decreases as altitude increases.

6. Which engine on a conventional twin-engine airplane is considered to be the critical engine? Why?

7. Define accelerate-stop distance and explain why it is an important performance consideration.

8. How would a CG location which is aft of limits affect single-engine operation of a conventional twin?

9. Is it a good operating practice to conduct engine-out operations below the *published* V_{MC} if the pilot knows the aircraft is significantly below gross weight while operating at high density altitudes?

10. What multi-engine airplanes are required to have a positive single-engine climb rate at 5,000 feet?

11. What is the significance of a blue line on the airspeed indicator of some multi-engine airplanes?

12. Explain the meaning of single-engine service ceiling and single-engine absolute ceiling.

13. Explain how the best single-engine rate-of-climb and best single-engine angle-of-climb airspeeds change with an increase in altitude.

14. If an engine failure occurs immediately after takeoff, which is more valuable—airspeed in excess of the single-engine best rate-of-climb speed or additional altitude? Why?

15. Explain why the climb performance of a twin-engine airplane is decreased more than 50 percent with one engine inoperative.

MULTI-ENGINE EMERGENCY PROCEDURES

1. Explain the techniques used to induce simulated engine failure during takeoff roll. At what airspeed should this emergency be simulated?

2. What is the safest method of simulating an engine failure during the initial climb, when the landing gear is up or in transit?

3. What methods can be used to simulate engine failure during cruise?

4. Assume a conventional twin-engine airplane is trimmed for flight with the right engine inoperative. What control problems may be encountered if power is applied rapidly on the right engine, and how can this reaction be prevented?

5. Although engine fires cannot be authentically simulated in flight, the student must learn the techniques used to meet this emergency. What are the items requiring immediate attention if an engine fire occurs?

6. How should an instructor introduce emergency situations such as electrical fires, flap malfunctions, and landing gear malfunctions?

7. What types of emergency situations require an emergency descent?

8. Explain why an applicant for a multi-engine rating should be discouraged from attempting single-engine go-arounds.

9. What methods should be used to identify the inoperative engine on a conventional twin-engine airplane?

10. Explain the procedure for isolating the source of an electrical fire.

11. Under what flight conditions should engine-out airwork be practiced with the inoperative engine shut down and the propeller feathered?

12. If smoke enters the cabin due to fire, what precautions should the pilot be aware of concerning opening the door or window to expel the smoke? What is the preferred method of smoke evacuation?

GENERAL SUBJECTS

1. List the factors which cause the greatest performance loss after an engine failure.

2. What should a student learn from an engine inoperative loss of directional control demonstration in a conventional twin?

3. Why should an engine failure during takeoff be simulated at an airspeed less than V_{MC}?

4. Why are single-engine rejected landings practiced only at a safe altitude?

5. What criteria should be used to evaluate an applicant's preparedness for a multi-engine practical test?

6. Explain the objective of demonstrating the effects of various airplane configurations on engine-out performance.

7. Explain why the engine starter must be used during an in-flight engine restart, if the propeller has been feathered.

8. What liftoff airspeed is recommended by the FAA for a maximum performance takeoff?

9. If an engine failure occurs above the airplane's single-engine absolute ceiling, what airspeed provides the slowest rate of descent?

10. Explain why takeoff and initial climb are the two most critical phases of flight in regard to an engine failure.

11. Why is the throttle of the suspected inoperative engine reduced to idle as the first step in the feathering sequence?

12. Explain why the left engine is considered to be the critical engine in terms of engine failure.

13. What atmospheric conditions might prevent performance of an engine inoperative loss of directional control demonstration?

14. Describe the correct method for performing imminent stalls in the gear-down and full-flaps configuration.

15. Why is it important to teach a student to calculate such single-engine performance data as accelerate-stop distance, single-engine takeoff distance, single-engine rate of climb, and single-engine service ceiling during preparations for a flight in a multi-engine airplane?

ANSWERS

EXERCISE 1A — SINGLE-ENGINE AIRPLANE INSTRUCTION

1. 4
2. stall avoidance
3. realistic distractions
4. True
5. pitch attitude
6. 3
7. True
8. wind
9. downwind
10. airspeed, approach angle
11. P-factor
12. 4
13. True
14. False
15. V_X, V_Y
16. False
17. pilotage, dead reckoning, radio navigation
18. 2
19. 2
20. False
21. False
22. True
23. decreases
24. 2
25. False

EXERCISE 1B — AERODYNAMICS OF FLIGHT

1. acceleration, deceleration
2. equilibrium
3. above, below
4. impact pressure
5. shape, angle of attack
6. chord line, relative wind
7. 2
8. 1
9. separate
10. washout
11. False
12. True
13. True
14. False
15. False
16. 2
17. 1
18. 1
19. 4
20. efficiency
21. 3
22. thrust
23. angle of attack
24. descending
25. yaw, left
26. angle of attack
27. True
28. retracted
29. horizontal
30. E
31. C
32. D
33. A
34. B

EXERCISE 2A — THE LEARNING PROCESS

1. True
2. exercise
3. intensity
4. True
5. 4
6. False
7. 1
8. insight
9. basic needs
10. False
11. correlation
12. False
13. A
14. B
15. C
16. B
17. 2
18. C
19. A
20. D
21. B

EXERCISE 2B — THE TEACHING PROCESS

1. C
2. B
3. D
4. A
5. True
6. False
7. False
8. habits
9. True
10. 2
11. motivated
12. 1. Introducing new material
    2. Summarizing procedures
    3. Relating theory to practice
    4. Reemphasizing main points
13. True
14. True
15. 1. Overhead
    2. Rhetorical
    3. Direct
    4. Reverse
    5. Relay
16. instruction
17. False
18. True

EXERCISE 2C – PLANNING AND ORGANIZING

1. 2
2. 2
3. False
4. True

5. C
6. E
7. A
8. D

9. F
10. B
11. 2
12. True

EXERCISE 3A – AUTHORIZED INSTRUCTION AND ENDORSEMENTS

1. 2
2. E
3. B
4. F
5. D
6. A

7. B
8. False
9. False
10. 35, 20, 15, 10
11. True
12. False

13. 3
14. True
15. False
16. 3
17. False

EXERCISE 3B – REGULATIONS (FAR PART 61, SUBPART G)

1. True
2. quizzing, testing
3. 24, 200
4. False
5. 8, 24

6. class, type
7. False
8. five
9. False
10. logbook

11. False
12. True
13. 24
14. True

EXERCISE 4A – BASIC INSTRUMENT INSTRUCTION

1. electrically, vacuum
2. False
3. False
4. 8°
5. True
6. True
7. True
8. 3
9. True

10. False
11. 1
12. 1. Altimeter
    2. Airspeed indicator
    3. Vertical speed
       indicator
13. pilot's operating
    handbook
14. higher

15. True
16. False
17. straight-and-level
18. deviation, magnetic dip
19. right
20. interpretation, control
21. 10
22. attitude indicator
23. True

EXERCISE 4B – INSTRUMENT PROCEDURES INSTRUCTION

1. False
2. False
3. False
4. standard instrument
   departures
5. 3
6. B
7. C
8. C
9. A
10. A
11. A
12. A

13. 4
14. False
15. 2
16. visibility minimum
17. True
18. one-half
19. True
20. True
21. heading indicator
22. True
23. localizer
24. heading corrections
25. glide slope

26. False
27. True
28. False
29. False
30. True
31. 2,000, 3
32. True
33. ETAs
34. False
35. instrument competency
    check
36. 4

EXERCISE 5A — MULTI-ENGINE AIRPLANE INSTRUCTION

1. cockpit
2. familiarization
3. differential
4. performance
5. False
6. power
7. rate-of-climb
8. False
9. True
10. 2

11. 2
12. True
13. steep, high
14. pilot's operating handbook
15. rate-of-climb
16. flaps, landing gear
17. altitude, power
18. True
19. airspeed, configuration
20. directional control

21. response
22. altitude
23. False
24. handling characteristics
25. True
26. False
27. True
28. inoperative engine
29. pitch attitude

EXERCISE 5B — AERODYNAMICS OF MULTI-ENGINE FLIGHT

1. True
2. True
3. False
4. True
5. False
6. 4

7. critical engine
8. left
9. V_{MC}
10. 2
11. True
12. 3

13. 4
14. 2
15. 2
16. True

NOTES

NOTES

CFI Renewal In The Convenience Of Your Own Home

Renew your CFI certificate at home—on YOUR schedule—using Jeppesen's comprehensive and flexible FAA-approved program.

You will receive eight specially produced, high quality videos and eight chapters of comprehensive text material. Our flexible enrollment allows you to enroll as close as 30 days prior to your expiration date. And...Jeppesen can now issue your temporary CFI certificate directly to you. What could be more convenient!

Jeppesen's CFI Renewal Program also provides excellent continuing education for those CFIs not needing a renewal, and for those who also want to build their CFI reference library with quality Jeppesen materials.

Plus, we offer a toll-free CFI Hotline so you can get instant answers to your training questions. Item Number JS200943 $195.00

New! Short-Term Renewal Ability!
In special circumstances, we may be able to offer the Jeppesen CFI Certificate Renewal Program within 30 days prior to your CFI Expiration Date.

New! Jeppesen Now Has The Authority To Issue Temporary CFI Certificates!
Jeppesen's CFI Renewal Program is the most widely recognized professional CFI Renewal Program used today.

Jeppesen's CFI Renewal Program fits YOUR SCHEDULE.

JeppGuide: Makes Getting Around An Airport Easy

JeppGuide includes:Airport diagrams • Detailed listing of FBO services • Arrival, departure, taxi procedures; navigation aids, radio frequencies, and more.And free with every JeppGuide is a regional FBO Directory—The Pilot's Yellow Pages. This quick reference guide includes key location information and phone numbers.

Convenient Coverage

■ Northeast Connecticut, Delaware, Maine, Maryland, New Hampshire, New Jersey, New York, Pennsylvania, Rhode Island, Vermont, Virginia, West Virginia and Washington D.C.
Item Number AUSRAD40 $29.95*

■ Southeast Alabama, Florida, Georgia, Mississippi NorthCarolina, South Carolina and Tennessee.
Item Number AUSSAD40 $29.95*

■ South Central Arkansas, Louisiana, New Mexico, Oklahoma and Texas.
Item Number AUSTAD40 $29.95*

■ Great Lakes Illinois, Indiana, Kentucky, Michigan, Ohio, Wisconsin.
Item Number AUSPAD40 $29.95*

■ Western Arizona, California, Idaho, Nevada, Oregon, Utah and Washington.
Item Number AUSFAD40 $29.95*

■ North Central Colorado, Kansas, Missouri, Minnesota, Montana, Nebraska, North Dakota, Iowa, South Dakota and Wyoming.
Item Number AUSNAD40 $29.95*

*Note: Price includes 1 year Revision Service

PN-1 Navigation Plotter The classic opaque background allows easy identification of WAC and Sectional chart scales. The scales are calibrated in statute and nautical miles.
(Size: 12 1⁄4"x 4")
Item Number JS526500 $5.95

PJ-1 Rotating Azimuth Route planning made simple. One twist of the azimuth allows you to measure the true course or to determine the magnetic course. The PJ-1 also includes scales for sectional terminal area and WAC charts in statute and nautical miles.
(Size: 3 1⁄4"x 13"; 3⅜" Diam.)
Item Number JS526501 (English) $9.95
Item Number JS524412 (Spanish) $10.95

Metal CSG Heavy duty metal construction assures long life, high accuracy and enduring quality. Solves low and high speed problems. Non-glare finish. Complete with instruction manual and carrying case.
Item Number JS514105 $27.95

Student CSG
Full function, full size. Perfect for the budget-conscious student. Constructed of rugged plas-
ticized composite material. It conforms to the instructions and examples in Jeppesen manuals,
workbooks and audiovisual programs.
Item Number JS514101 $8.95

Private Pilot Exercise Package Covers material in the Private Pilot Manual. Includes
a series of comprehensive student exercises designed for individual study and self-check-
ing. Completion of the Exercise Book is required by the completion standards of FAR Part
141-approved schools. Includes sample Dallas-Ft. Worth sectional chart. 176 pages.
Item Number JS324772 $10.45

Private Pilot Maneuvers Manual Using over 80 figures, diagrams and illustrations, this
manual describes and illustrates maneuvers and procedures. 176 pages.
Item Number JS314705 $9.95

FAR/AIM Manual Comes complete with: Exclusive FAR study guide • FREE Update
Summary • Complete Controller Glossary • FAR Parts 1, 61, 67, 71, 73, 91 ,97, 135, 141,
HMR175 and NTSB 830 • Plus, much, much more.
Item Number JS314107 $11.95

FAA Written Exam Study Guides Contain FAA airplane questions arranged in the same
sequence as the chapters in our textbooks. Explanations are placed next to each question
and include study references to the page in our textbooks where the topic is covered.

Private Pilot Item Number JS312400 $12.95
Instrument Rating Item Number JS312401 $16.95
Commercial Pilot Item Number JS312402 $14.95

Private PTS Study Guide Prepares students for both the knowledge and the skill areas
of the Private Pilot, Airplane Single Engine Land, Practical Test Standards.
Item Number JS312404 $16.95

Professional Pilot Logbook The book the pros use. Aviation's most popular professional
logbook can handle 10 years of data. It includes simplified pilot and aircraft annual
summaries. . (Size: $6\frac{3}{4}$"x$11\frac{1}{4}$")
Item Number JS506050 $19.95

Pilot Logbook Designed for lifetime durability, this 96-page permanent record offers large
flight time category spaces, arranged according to FAA regulations. It's become the standard of the industry. (Size: $8\frac{5}{16}$"x$5\frac{5}{16}$")
Item Number JS506048 $7.95

FARs Explained (135) This book reviews the regulations for FAR Part 135, and looks
at FAA Advisory Circulars, the *Airman's Information Manual,* NTSB decisions, FAA
Chief Counsel opinions, and regulatory background to help pilots better understand the
regulations.
Item Number JS319013A $26.95

FARs Explained (61, 91, and NTSB 830) This Book has quickly become an industry
standard. It includes: Easy to read and understand explanations of FAA regulations
• Cross-references to aid understanding • Actual case histories and FAA Chief Counsel
Opinions.
Item Number JS319012 $26.95

No Matter How High Or How Far You Fly, It's Easier With TechStar
No other handheld electronic computer gives you as much capability. Pre-flight and inflight
planning couldn't be easier.

TechStar offers:

▲Over 25 Aviation Calculations
▲55 Conversions Available
▲6 Independent Memory Registers
▲Enter Values In Any Order, Or Change
  Them At Any Time
▲Pilot Friendly Use

▲Work On One Problem While Another On-
  Screen Calculation Is Being Completed
▲Weight And Balance Calculations
  Include Weight Shift Formulas
▲Three-Year Warranty
▲Batteries Included

TechStar Item Number JS506000 $65.95

FlighTime Video Series
Your Key to 60 Years of Aviation Experience

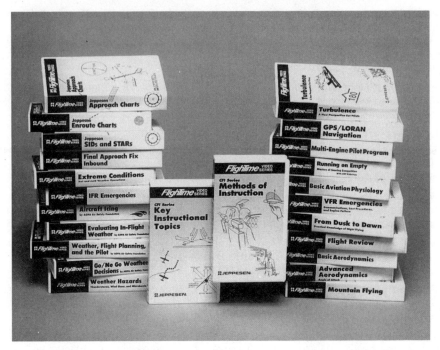

Instrument Flying

Title	
Jeppesen Chart Series — 3 Videos	$97.95
Approach Charts	36.95
Enroute Charts	36.95
SIDs and STARs	29.95
IFR Emergencies	36.95
Final Approach Fix Inbound	36.95

General Subjects

Title	
Mountain Flying	$29.95
Basic Aerodynamics	36.95
Advanced Aerodynamics — Angle of Attack	36.95
Flight Review	29.95
VFR Emergencies	36.95
From Dusk to Dawn	36.95
Basic Aviation Physiology	36.95
GPS/Loran	36.95

Weather Topics

Title	
Weather Hazards	$36.95
Extreme Conditions: Hot & Cold Weather	36.95
Aircraft Icing	29.95
Evaluating In-Flight Weather	29.95
Weather, Flight Planning & the Pilot	29.95
Go/No-Go Weather Decisions	29.95
Turbulence, A New Perspective	29.95

Special Topics

Title	
Multi-Engine Program	$44.95
Running on Empty - (Soaring)	29.95

New CFI Video Series

Title	
Key Instructional Topics	$36.95
Methods of Instruction	36.95

Visit Your Jeppesen Dealer or Call 1-800-621-JEPP
Prices subject to change